Infrastructure for the Built Environment

In loving memory of Rod Howes, a devoted husband and father,
who gave us all so much and was tragically taken from us
long before his time and shortly before the publication of this book.
We will remember him for his endless support and guidance in
helping us achieve our goals and aspirations and
we continue to be proud of his many accomplishments.

Infrastructure for the Built Environment

Global Procurement Strategies

Rodney Howes and Herbert Robinson

ELSEVIER

AMSTERDAM • BOSTON • HEIDELBERG • LONDON • NEW YORK • OXFORD
PARIS • SAN DIEGO • SAN FRANCISCO • SINGAPORE • SYDNEY • TOKYO

Butterworth-Heinemann is an imprint of Elsevier

Butterworth-Heinemann is an imprint of Elsevier
Linacre House, Jordan Hill, Oxford OX2 8DP
30 Corporate Drive, Suite 400, Burlington, MA 01803

First published 2005

British Library Cataloguing in Publication Data
A catalogue record for this book is available from the British Library

Library of Congress Cataloguing in Publication Data
A catalogue record for this book is available from the Library of
Congress

ISBN-13: 978-0-7506-6870-5
ISBN-10: 0-7506-6870-9

05 06 07 08 09 10   10 9 8 7 6 5 4 3 2 1

For information on all Butterworth-Heinemann publications
visit our website at http://books.elsevier.com

Typeset by Charon Tec Pvt. Ltd, Chennai, India
www.charontec.com
Printed and bound in Great Britain

Contents

List of case studies

List of figures

List of tables

Acknowledgement

Professor Rodney Howes (1943–2005)

This book is dedicated to the memory of Professor Rodney Howes, known to his friends and colleagues as Rod, and to his students as 'Prof'. He made significant contributions to the development of the construction management discipline and was held in high esteem by his peers in academia and the construction industry both nationally and internationally. For more than three decades, he was the powerhouse of thinking and academic leadership in the development of the School of Construction at London South Bank University from its origins in the Brixton School of Building.

Rod started his career in the construction industry as a planning engineer rising to senior planner in a major UK construction company. He gained professional recognition and was admitted as a member of the Chartered Institute of Building (CIOB) in 1964. He joined academia in 1967 as a lecturer in the renowned Brixton School of Building. Here his industrial experience was effectively utilised in the development and delivery of various building and engineering courses and he became a member of the British Institute of Management in 1968. His quest for advanced knowledge and academic excellence led him to enrol in a research programme at the Department of Building Technology and was awarded a PhD from Brunel University in 1983 – a department noted at the time for its interdisciplinary approach to building engineering. It was his blend of industry and research experience which acted as a catalyst in developing and updating courses to constantly challenge students and to reflect the changing needs of industry. In many ways, Rod was an inspirational leader with a clear vision and determination to achieve higher standards.

Following the significant challenges facing higher education at the time, the Brixton School of Building evolved to become part of the newly established South Bank Polytechnic in 1971. Rod was promoted to a senior lectureship in 1971 and rose to the position of Head of the Department of Building Administration which later became the Department of Construction Management in 1986. He provided visionary and strategic leadership which led to significant growth in teaching, research, and academic enterprise activities. He had long recognised the importance of research underpinning teaching and was ahead of his peers in the Polytechnic in attracting research grant funding from prestigious sources such as the UK Research Councils. At the time, such funds were the preserve of research-led universities. He undertook original research in project management systems in collaboration with industry, leading to the deployment and use of software solutions in practice. He was promoted to the Chair and Professor of Construction Management in 1990 in recognition of his outstanding contribution to teaching, research, enterprise, administration and academic leadership. He was an academic magnet in recruiting, developing and retaining staff with outstanding research potential. When the Polytechnic was granted University status in 1992 his department was already ahead of the rest in its portfolio of research and enterprise activities. He became the Head of School of Construction Economics and Management in the same year following a merger. He maintained his headship of the much expanded School of Construction in 1995, which subsumed Civil Engineering in the Faculty of the Built Environment at South Bank University. He was instrumental in introducing innovative academic management practices in the development of the School and building collaborative links with industry and universities in the UK and abroad.

Under his leadership, the department was highly successful in attracting significant numbers of students from around the world, particularly from South East Asia, Africa, and Europe, generating substantial funds to further develop the School. He oversaw the creation of research groups operating in key areas of construction management, international economics, water engineering, geotechnics, civil engineering management, housing and urban regeneration. He attracted and recruited a growing number of research active staff and supervised PhD students to support the School's research strategy, many of whom have gone on to achieve prominent positions as academics and professors in leading universities in the UK and abroad. Whilst at London South Bank University he also developed strong links with universities in Europe and extensive research contacts with universities in mainland China and Hong Kong. As an education consultant, Rod played a leading role in the development of courses and the establishment of the first University in The Gambia.

Rod was extensively involved outside the University as external examiner for undergraduate and postgraduate courses and research degrees at many universities in the UK and abroad. He made significant contributions to the activities of professional bodies and associations related to the built environment. He was co-founder and inaugural Chairman of the Association of Researchers in Construction Management (ARCOM), Chairman of the Council of Professors in Building Engineering and Management, Construction Industry Council's (CIC) Innovation and Research Committee, Innovation and Research Task Force (CIOB) and the Quality Assurance Agency (QAA) Building and Surveying Benchmarking Group. He was also actively involved in other organisations, including the UK Construction Research and Innovation Strategy Panel (CRISP), the Advisory Committee of CIRIA's Construction Management and IT Research Programme and various other scientific committees for international conferences. He has published numerous articles and acted as a referee for major research councils in the UK (EPSRC, RSRC, Royal Society) and abroad, and for leading international journals in the Built Environment.

Rod retired as a member of the Senior Management Team in the Faculty of the Built Environment at London South Bank University in 2001 to become an Emeritus Professor. This allowed him to focus and concentrate on things that he was unable to do whilst in a full-time academic post. Not surprisingly, his book (co-authored with Professor Tah at Salford University) entitled 'Strategic Management Applied to International Construction' was published by Thomas Telford shortly after in 2003. He continued to serve academia and industry in various capacities as an ECCREDI Council Member, member of the Chartered Institute of Building (CIOB) Innovation and Research Panel and the Foundation for the Built Environment, academic referee for international journals and research councils. He was a Visiting Scholar at Hong Kong Polytechnic University and was also recently appointed as a Visiting Professor at Salford University.

Rod's industrial and academic experience spanned more than 40 years. During this period, he has made immense contributions to academia and the construction industry both nationally and internationally. For those who knew Rod well, he was kind, courteous, respected and extremely hardworking. Indeed, he worked until the very last day. He was tragically killed in a cycling accident on 30th August 2005, the day after putting the finishing touches to this book, during his morning cycle route through the Kent country lanes. The unbelievable and devastating news of his death shocked all his family, friends and colleagues. This book will therefore serve as a fitting and lasting memorial for the life and achievements of Professor Rodney Howes. He will be greatly missed by members of the academic community and professional organisations in the built environment as a colleague, teacher, friend, mentor and advocate.

Joe Tah (Professor and Associate Head of School, Salford University)
Barry Symonds (Head of Department, London South Bank University)
Herbert Robinson (Senior Lecturer, London South Bank University)

Foreword

Physical infrastructure is central to socio-economic development at national, regional and global levels. Investment in physical infrastructure has traditionally been the responsibility of government, and funding has been provided from taxation and the issue of government bonds. Despite this there is a long track record of private investment in infrastructure typified by land owners charging tolls for passage over their domain and businesses that have invested in infrastructure to support trade and commerce, as in the case of Trans American railroads. More recently, private investment in public infrastructure has been encouraged by governments worldwide in accordance with the concept of public private partnerships. This development has opened up additional business opportunities for the suppliers of physical infrastructure to provide facility management services extending over concessionary periods within the operational life cycle.

Infrastructure in the form of transport, power generation and distribution, water services, waste disposal, telecommunications, health, education, public and commercial buildings has become the life blood of urban and rural communities. Cities have become extremely complex and are highly vulnerable to disruption caused by physical infrastructure failure. For example, recent power failures in London and New York, caused by a combination of circumstances resulted in system overloads that created enormous disruption and a threat to public safety. It has been argued that a lack of investment was to blame, which is partially true. However, these events have served to highlight how fragile and dependent cities have become on adequate and reliable physical infrastructure. Furthermore, an enormous appetite has been developed for energy and other resources, together with the need for advanced technology and management systems to maintain infrastructure services at levels expected by the public.

There is a growing realisation that increasing demand for non-replaceable resources to support economic growth cannot be maintained indefinitely, especially taking into account the amount of waste generated and current levels of conservation. In the long-term existing urban conurbations are not sustainable, since the current non-replaceable resource usage rate will eventually result in complete depletion. There is therefore an urgent need to reduce waste and to conserve resources by recycling and measures must be taken to reduce energy consumption by means of efficiency gains. The continued burning of fossil fuels is increasing pollution by the release of carbon and greenhouse gases into the atmosphere. This is contributing to global warming and climatic disruption resulting in rising sea levels, flooding and damage caused by high winds. Increasingly, infrastructure is being subjected to more demanding conditions and standards of structural and enclosure performance.

The research required to write this book has underlined the importance of taking a long-term holistic view of the demand and supply of physical infrastructure necessary to fulfil economic, social and cultural aspirations. Expectations and standards currently demanded by society require that cities and towns operate as integrated systems with all parts performing in unison to achieve a comfortable and safe living environment where individuals, organisations and institutions can enjoy prosperity and well-being. Hence this book concentrates on bringing together the different aspects of physical infrastructure delivery and covers a broad range of subject areas concerned with infrastructure economics and management. The aim is to provide an integrated and holistic approach to enable understanding from the formulation of policy to implementation, operation and conservation of sustainable infrastructure. Many of the ideas contained in this book are based on system thinking, and readers are encouraged to view physical infrastructure as a vital organ that sustains the way modern society operates and develops to achieve human dreams and visions for the future.

The principles and theories described in each chapter have addressed issues across national boundaries and due recognition has been given to the differing requirements of advanced industrial, developing and least developed countries. The case studies cited are intended to reinforce theory and to provide a practical dimension to each chapter.

This book is intended to provide knowledge and understanding to a wide range of individuals, organisations and institutions and will be an essential text, particularly for policy makers, practitioners, academics, undergraduate and postgraduate students. The international dimension, together with the citation of important case studies in different countries, will appeal to a worldwide readership.

Numerous illustrations and tables support the text, and references have been provided for readers who wish to expand their knowledge in a particular area covered by the book.

Rod Howes and Herbert Robinson

Chapter 1

Introduction

Overview

Infrastructure is central to the socio-economic development of all countries and the well-being and prosperity of society. Increased globalisation means that the level and quality of infrastructure development has become more critical than ever for national economies to remain competitive and innovative in the global market place.

The term 'infrastructure' is generic and it can be interpreted broadly as physical, personal and institutional. This book focuses on physical infrastructure provided in the form of civil engineering and building projects. Broadly speaking, these works concern economic infrastructure such as transportation, energy, water, telecommunications and the provision of trade and social infrastructure, specifically public administration, commercial, industrial, healthcare, education and residential buildings.

However, infrastructure development is a complex process influenced by a variety of policy and strategic factors. It is also the contextual variables of a country such as economic, social, environmental and political forces, which interact to determine the composition, priorities and timing of infrastructure programmes. Increased globalisation means greater interconnectivity at the levels of individuals, interest groups, businesses, interaction between countries, national and international development agencies and a significant growth of international trade. This has increased the urgency for infrastructure development in developing as well as developed countries.

Developing appropriate policy and strategic frameworks are therefore fundamental to infrastructure delivery, so is the need for practical approaches to facilitate the implementation of infrastructure projects and sustaining the services derived from their use. Over the past years, numerous studies have been carried out on various aspects of infrastructure delivery such as infrastructure policy, planning, design, construction and operations. But these findings have not often been presented in an integrated way to facilitate the understanding of different components in the infrastructure delivery chain. This book therefore focuses on bringing together the different aspects of infrastructure delivery together and covers a broad range of subject areas in infrastructure economics and management. The aim is to provide a holistic approach to facilitate understanding of the interrelationship between key elements of the infrastructure delivery chain from policy formulation to operations, conservation and sustainability of infrastructure.

Most of the ideas contained in this book are underpinned by systems thinking encouraging readers to view infrastructure in a holistic rather than the traditional often fragmented way. An important contribution of the book is therefore the emphasis on interconnectivity encapsulated in the concept of the infrastructure delivery chain and more importantly the increased emphasis on services, which is the ultimate objective of infrastructure projects. It is this new approach that makes this book different from other mainstream textbooks on infrastructure planning, design, construction and facilities management. The book should therefore appeal to a wide range of people and will be an essential reading particularly for policy makers, practitioners, academics, undergraduate and postgraduate students involved in any aspect of infrastructure delivery. The concepts, theories and principles described in each chapter are generally applicable across national boundaries and are supplemented by detailed case studies and commentary that provide examples of practical applications and lessons learned. References are also provided at the end of the chapters for further study and detailed investigations on important aspects are discussed.

The book has 11 chapters organised in three parts to reflect the different aspects of the infrastructure delivery chain as shown in Figure 1.1.

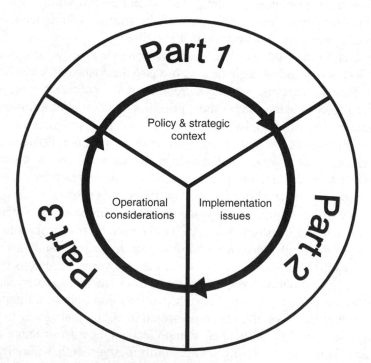

Figure 1.1   Infrastructure delivery chain.

The first part of the book (Chapters 2–4) focuses on the policy and strategic context to provide the foundation and the building blocks for infrastructure delivery. The key themes are nature of infrastructure, policy selection and strategic options. The second part (Chapters 5–8) addresses the practical aspects of policies and strategies relating to the evaluation and implementation of infrastructure projects and discusses issues such as the selection of the most appropriate projects and associated resource implications, procurement choices and the financing mechanisms. The key themes are project evaluation, procurement choices and funding options. The third part of the book (Chapters 9–10) covers the operational aspects of infrastructure, which is arguably the most important part but is often neglected. Issues discussed related to sustainable design, conservation and facilities management.

Part 1: Policy and strategic context

The nature of infrastructure projects

Chapter 2 sets the scene for the subsequent chapters (3 and 4) by briefly presenting the concept of infrastructure from different perspectives, highlighting the need for a holistic approach in defining infrastructure to inform the development of policy and strategy. Using Jochimsen's approach the different components of infrastructure (personal, physical and institutional) are outlined and their role in service delivery explained to understand the development of policy.

The classification of different types of physical infrastructure is discussed in terms of their functional characteristics and morphological features as they have important policy implications. Within the global context of infrastructure development, two basic components of infrastructure – provision and production and the development stages (planning, design, construction, operation, disposal and recycling) are explained. The need for a serviced-focussed approach is discussed as crucial in sustaining infrastructure projects in the light of recent procurement options such as public–private partnership to improve the level of infrastructure services.

Determination and selection of infrastructure policy

Chapter 3 builds on the previous chapter by providing the context for developing a policy framework to facilitate the development of infrastructure projects. This chapter examines the policy context for the delivery of infrastructure. Central to the policy making process are the institutions, goals and objectives, resources, knowledge, information and communication systems. The environment is also crucial as it reflects the economic, social and political context that influence the policy making process. The formulation and implementation of appropriate infrastructure policies therefore require an understanding of these key elements and their interactions.

The factors necessary to develop or strengthen policy capacity, to improve its effectiveness to continuously attract investment and to deliver sustainable infrastructure services are explored.

The chapter also discusses the key criteria that determine public or private participation in the delivery of infrastructure services. It is argued that decisions on whether the private or public sector should provide a particular type of infrastructure are rarely based on economic considerations alone; hence a range of environmental, social and political factors come into play such as public service obligations, externalities (third party costs and benefits), national security and defence.

Strategic considerations for infrastructure development

Chapter 4 explores the strategic implications of the policy issues discussed in the previous chapter by focusing on the three major thrusts for the development of an infrastructure delivery strategy – the need for investment, identifying knowledge resources and gaps, and the role of a planning and regulatory framework.

The complexity of the infrastructure development process and the need for a holistic approach to identify systemic failures in the infrastructure delivery chain is discussed using a variety of systems diagramming techniques. It is argued that the contextual variables of a country such as economic, social, environmental and political forces interact to determine the composition, priorities and timing of infrastructure programmes. These variables have a significant impact on the availability of materials and skilled labour, balance of payments, employment, imports and exports, pollution, private sector involvement, and land use policies which affects the infrastructure delivery chain. Some of these variables are internally driven whilst others are externally imposed.

The chapter emphasizes the need for a service-focused approach to strengthen the infrastructure delivery chain underpinned by the concept of value added to identify bottlenecks and to continuously improve services. The need for a decision making hierarchy to facilitate the management of resource demand and supply to improve infrastructure services is also explored.

Part 2: Implementation issues

Project evaluation and resources

Chapter 5 addresses evaluation issues focusing on how projects are selected based on macro variables reflecting national priorities and micro variables at project level. The chapter discusses the key contextual (macro) variables such as economic, social, environmental and institutional factors, and project specific (micro) variables such as design choice, construction methods, size and type of infrastructure development.

The need for two levels of evaluation is explained. Macro evaluation is undertaken at the national level by government planners, policy and decision makers focusing on the overall goal of a project in terms of its broad impact on economic, socio-cultural, institutional, political and environmental indicators. Micro evaluation at the project level by the project owner focuses on issues of immediate concern to the project in terms of its viability with respect to specific objectives relating to the type of development, design and construction options. It is therefore argued that a holistic and dynamic view is required in evaluating a project to reflect both project (micro) level and national (macro) level considerations.

The chapter also provides a detailed analyses of the resource implications of infrastructure projects and discusses the need for the management of resources to improve the delivery of infrastructure services. The underlying principles, strengths and weaknesses of different types of resource management models are outlined and a case is made for an InfORMED framework to ensure that a country's infrastructure development strategy is implemented in such a way that it reflects the resource context and policy priorities. This is particularly crucial in developing countries where resource markets are often underdeveloped.

Infrastructure project procurement

Chapter 6 discusses how projects are implemented using various procurement options. The selection of the most appropriate procurement option for the provision of physical infrastructure is a key action that contributes to project success by satisfying client and user needs. Important decisions are required concerning the extent of public control necessary and the desirability to leverage in private investment by the provision of business opportunities and incentives. Chapter 6 identifies the elements of a procurement strategy and proposes a framework for the identification and categorisation of procurement methods according to the nature, complexity, scale and environment within which the project will be realised. Recognition is given to the need for innovation and research into products and processes to increase competitive advantage and to improve efficiency and sustainability over the whole project life cycle.

Specific coverage is given to public–private partnerships (PPPs) and the implications brought about by concessionary agreements that incorporate service levels and performance benchmarks. The private finance initiative (PFI) originally developed and implemented in the UK is described.

A portfolio-based strategy is proposed and key elements are identified to ensure that projects brought forward for procurement are in accordance with policy and strategies that deliver physical infrastructure according to laid down priorities.

Additional considerations for infrastructure procurement in developing countries are explained and available assistance is identified in the form of loans and grants from the World Bank, regional development banks and aid agencies. Private investment by means of the World Bank's model of private participation in infrastructure (PPI) is explained and the selection of the most appropriate PPI option is considered.

The chapter concludes with guidance regarding the selection of bidders and the evaluation of bids.

The financing of infrastructure projects

Chapter 7 focuses on financing infrastructure projects within the context of affordability, feasibility and sustainability. Various financing mechanisms are discussed, namely public financing supported by taxation and public borrowing and private financing supported by private investment. These are combined into a dual track strategy leading to PPP. In the context of private financing or ownership, the need for regulation of physical infrastructure in the public interest is addressed and the functions of the regulator or regulating body are described briefly.

Chapter 7 concentrates on project analysis, recognising the importance of project costs, revenues or benefits, return on investment and the identification, assessment and evaluation of risk. Discounted cash flow (DCF) and net present value (NPV) are explained and the concept of scenario comparison is introduced. Uncertainty associated with NPVs is used to assess risk and stochastic decision trees are demonstrated as a means of assessing probability and chance associated with strategic decision making.

The financial appraisal of bidders is considered using project appraisal criteria and weightings. In this manner project objectives can be evaluated and different project proposals assessed regarding their potential benefits.

The need for adequate project and portfolio financial control is explained with specific reference to the control of the construction process and the monitoring and control of operational costs post occupation.

Funding infrastructure projects in developing countries

Chapter 8 discusses the special funding needs and requirements for infrastructure provision in developing countries. The provision of essential physical infrastructure is necessary to maintain public health and to support economic growth. Developing countries have enormous requirements for infrastructure development in order to reduce poverty and improve health and advance living standards. The roles of the World Bank, regional development banks and aid agencies are described and profiled. The procedure for loan applications is outlined and opportunities for greater involvement of private finance to support new infrastructure projects are explored.

The potential for co-financing physical infrastructure projects is investigated and conflicts of interest are highlighted. Other considerations concerning lending conditions, cost sharing and foreign currency constraints, the effect of cross default and other conditional clauses are explained.

Emphasis is given to adequate physical infrastructure project identification and the development of capacity to improve efficiency and to develop more robust sustainable solutions. A balanced approach is proposed that takes into account the need for technology transfer, but not at the expense of utilising local resources, including the creation of local jobs, together with the development of skills and expertise.

Chapter 8 concludes by examining three detailed case studies that have taken advantage of co-financing arrangements and have engaged adequately with the local community and its beneficiaries at every stage.

Part 3: Operational considerations

Towards sustainable infrastructure and conservation

Chapter 9 begins by explaining the guiding principles derived from earth summits and then suggests approaches as to how these might be best implemented to improve the human living environment and the prosperity of nations and their citizens. Consideration is given to the main drivers that influence the development of a strategy for the design and construction of infrastructure in support of sustainable development and conservation. Sustainable development is dependent upon design and construction decisions reflecting the conservation of non-replaceable resources and greater efficiency brought about by improved technology, waste reduction and recycling. The design and construction of physical infrastructure therefore has a key role to play in improving life cycle performance by expediting space utilisation and energy consumption and maximising replaceable resources, e.g., solar energy and natural advantages associated with site selection, orientation, use of gravity and passive ventilation.

Sustainable development is defined and the Triple Link Sustainability model is further conceptualised by the introduction of a Five Capital Model as proposed by Serageldin and Stee (1994). A justification is provided for sustainable construction (Parkin et al., 2003) and a strategy for environmental sustainability and conservation is proposed based on a national, regional and local planning framework.

Evidence is provided concerning the effect that greenhouse gases (GHGs) are having on climate change and global warming. Steps are also cited towards developing a low carbon economy and suggestions are made for new approaches and technologies aimed at reducing energy consumption and placing greater reliance on power generation from renewable resources. An assessment is made of the prime categories of physical infrastructure and recommendations are made for the future development of transport, energy generation, water systems and waste disposal crucial for the functioning of many types of economic, social and trade infrastructure.

A framework is provided as a basis for strategic environmental assessment using social, environmental, economic and institutional themes. These are further developed into sub-themes and assessment indicators. Chapter 9 concludes with a description of various impact assessment methodologies.

Facility management of infrastructure

Chapter 10 describes the theory of facility management and suggests how it should be applied to the operation of infrastructure. The definition and scope of the facility management of physical infrastructure is discussed. Reference is made to the growing popularity of PPP and the need for a holistic and integrated approach that extends from project inception to cover the whole project life cycle. The components of infrastructure facility management are identified and related to each other to give support for the case of facility management as an essential function.

The professional skills required by the facility manager are described and the importance of interaction between the design team and the facility manager to elicit operational knowledge and experience is stressed. Design quality indicators (DQIs) are cited as a means of measuring the performance of the overall design, together with elements and components thereof covering the whole life cycle.

A strategic approach to facility management is considered in relation to the categorisation of running costs and the need for value analysis to determine optimal solutions. The options for operational facility management are evaluated with reference to the consideration of in-house provision, outsourcing and a combination of both. The role of the client is examined, together with the packaging of outsourced contracts and contractual arrangements. The chapter concludes with an appraisal of monitoring, management, control and costing systems.

Conclusion and future horizons

The book concludes with Chapter 11 by synthesizing the main issues from all the chapters and providing a prognosis. This chapter takes a forward view relating to the adoption of a policy framework and proposes a systemic approach to its implementation. The sustainable use of resources, including renewable resources is examined and views are expressed on how future physical infrastructure provision should be developed and implemented.

The principles associated with a physical infrastructure policy framework are described and the influences on policy selection are identified, together with the factors that affect the demand. The development of strategy is shown to have its roots in policy, hence demonstrating the need for harmony and integration. A systemic and holistic approach is taken to the provision of physical infrastructure at the macro level and a Viable Systems model is proposed for use at the national and regional level in conjunction with the InfORMED System Model.

The case for sustainable infrastructure is supported by the need to develop and implement conservation measures and the development of new technologies and energy sources. It is suggested that the only major development currently on the horizon with the potential provide for growing energy requirements is hydrogen technology. The renewable energy sources cited, when considered individually, are relatively small scale by comparison and in many instances, as has been previously stated, are very dependent on favourable year round local conditions.

Taking a more positive attitude, it would appear that energy produced from water, wind, solar and biomass sources collectively has the potential to create substantial amounts of electricity. This depends on the correct selection of generation that suits local conditions and the development of technologies capable of extracting more power output from renewable sources. Hence, there is a case for investing in future R&D to unlock the potential for sustainable energy production.

The introduction of PPPs has helped to raise public expectation of amenity and services received from new and improved infrastructure. The impact of PPP is immense when one considers the scale of take up across the world and it is now established as a viable means of providing hitherto unaffordable infrastructure. It is now appropriate to take account of what has been achieved and what should be done to seek continuous improvement in the future. The next step proposed is to improve the successful delivery of infrastructure projects and to accelerate the process of procurement, which at the moment takes too long. The imperative is identified as the need to obtain the best possible service from infrastructure investment that has the capability to remove impediments to productivity and business enterprise.

Innovative methods for financing physical infrastructure are proposed using both public and private financing separately and in combination. The aim is to remove impediments and barriers in order to ease financial pressures and to make more efficient use of equity and debit capital available.

The chapter concludes with an appraisal of the importance of risk management throughout the whole life cycle of infrastructure projects and it identifies the need for close liaison and communication between those assessing risk and project managers.

Summary

This book provides an overview into the issues that influence the global provision of physical infrastructure. It has attempted to logically develop principles, theories and experience and then relate these to practice with the assistance of case studies at the end of each chapter. The initial chapters have concentrated on defining the nature of infrastructure prior to explaining developing concepts relating to policy and strategy. The realisation of physical infrastructure projects is covered by chapters concerning procurement and finance. Attention has also been given to the important issue of sustainability followed by an appraisal of facility management covering operational aspects. The final chapter takes a forward view and makes suggestions and recommendations concerning the future development and provision of physical infrastructure.

References

Atkin, B. and Brooks, A. (2000), '*Total facilities management*', Blackwell Science, London.

Brundtland, G.H. (1987), '*Our common future, World commission on the environment and development*', Oxford Paperbacks.

Checkland, P.B. (1981), '*Systems thinking, systems practice*', Wiley, New York.

Farrall, S. (1999), '*Financing transport infrastructure: policies and practice in Western Europe*', MacMillan, London.

Flood, R.L. and Jackson, M.C. (1991), '*Creative problem solving: total systems intervention*', Wiley, New York.

International Finance Corporation (IFC) (1996), '*Financing private infrastructure, lessons of experience*', World Bank, Washington, D.C.

Levy, S.M. (1996), '*BOT: paving the way for tomorrow's infrastructure*', Wiley, New York.

Parkin, S., Sommer, F. and Uren, S. (2003), '*Sustainable development: understanding the concept and practical challenge*', Proceedings of the Institution of Civil Engineers 156, March Issue ESI, pp. 19–26.

Miller, J.B. (2000), '*Principles and practice of public and private infrastructure delivery*', Kluwer Academic Publications, Dordrecht.

NZIER (2004), '*Sustainable infrastructure: a policy framework*', Report to the Ministry of Economic Development, New Zealand.

Part 1

Policy & strategic context

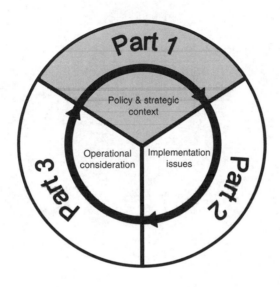

Key themes

nature of infrastructure
policy selection
strategic options

Chapter 2

The nature of infrastructure projects

Introduction

Infrastructure services are central to household, community and economic activities and they are important to improve public services, facilitate human development, economic growth and productivity in industry. This is reflected in the efforts of governments throughout the world to continuously increase investment. As countries aspire to higher levels of socio-economic and human development, the need to meet the increasing demand for infrastructure has become critical. The growing demand is due to changes in household activities, community and social changes such as work–home–leisure patterns, industrial needs and increased global interactions between countries. Failure to respond to demand will cause bottlenecks to economic growth and stifle human development.

Infrastructure development is increasingly a global business and complex activity involving the interaction of national and international agencies as a result of a number of factors. First, increased budgetary constraints for investment and maintenance of infrastructure faced by governments all over the world has increased the need for funding from international and bilateral development agencies and private investment from international investors and multinational companies. Certain infrastructure services, which traditionally are the responsibility of governments, have been open up to the private sector for investment. Second, the growth of international trade has provided wider access to the infrastructure market. For example, the new internet portal (Iworld) was set up to develop global infrastructure markets by linking buyers and sellers of infrastructure development and services around the world. Road, sea and air transport, power, water, telecommunication facilities and security infrastructure are essential services for multinational companies operating globally to maximise the benefits of their operations. Equally important are healthcare, education, and other social service provision central to the development of healthy and dynamic citizens, and a productive and knowledge creating workforce. Third, changes in international security and the relationships between nation states caused by terrorism, drug and human trafficking have implications for certain infrastructure that has been traditionally considered as the sole responsibility of national governments. There is now a far greater concern about security in airports, air traffic systems, seaports, other transport and communication infrastructure than ever before and the implication is that certain types of infrastructure should no longer be considered as national goods but 'global goods' as their impact is beyond national boundaries.

This chapter discusses the concept of infrastructure illustrating it with a number of definitions reflecting various perspectives, the need for a holistic approach and the implication for policy making. It presents a classification of physical infrastructure based on functional and morphological features. Further, it discusses infrastructure demand in a global context, the key aspects of infrastructure development, the infrastructure project cycle/stages, the key issues and actors involved from planning to recycling and disposal. This chapter also focuses on the need for a service-focussed approach necessary for the development of infrastructure projects.

Concept of 'infrastructure'

The term 'infrastructure' is often ambiguous as it is widely used in different context. In every day usage, it tends to be used to refer to a wide range of things from military installations, information technology, buildings to physical networks such as transportation and water systems. Development economists often refer to infrastructure as 'social overhead capital' described as investments in networks such as transportation, water and sewerage, power, communication and irrigation systems. From an economic standpoint, the production of such networked facilities are capital-intensive, traditionally owned and managed by the public sector; hence they are sometimes called public infrastructure capital. However, power, communication, highways and other infrastructure facilities can now be privately owned and managed. The implication is that this narrow definition of infrastructure focusing on public infrastructure capital which is often referred to as 'social overhead' is problematic since it only represents part of infrastructure provision. There are a variety of definitions for infrastructure from an industrial or national perspective. Sloman (1991) defined infrastructure as 'the facilities, support services, skills, and experience that supports a particular industry'. The New Collins Dictionary and Thesaurus defined infrastructure as 'the basic structure of an organisation, systems etc., or the stock of fixed capital equipment in a country including factories, roads, schools etc. considered as a determinant of economic growth'. Miller (2000) defined the term 'infrastructure' in a broad sense to mean collectively: capital facilities such as buildings, housing, factories and other structures which provide shelter; the transportation of people, goods and information; and the provision of public services and utilities such as water, waste removal and environmental restoration. These variations illustrate the considerable difficulties in trying to understand the concept of infrastructure, and its operationalisation for policy making. It also reflects the need for a holistic approach in defining the concept of infrastructure. The most elaborate attempt at a systematic definition was provided by Jochimsen (1966) who defined infrastructure as 'the sum of all basic material structures, institutional conditions and human resources available to society, needed for the proper functioning of the economic sector'. Jochimsen further distinguished between three components of infrastructure that are interrelated – institutional infrastructure, personal infrastructure and physical infrastructure (see Figure 2.1).

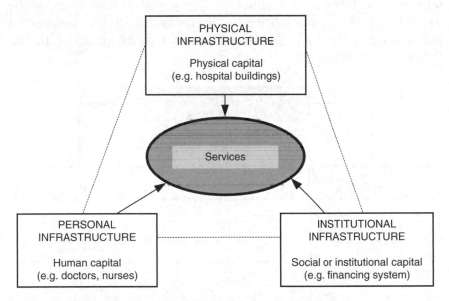

Figure 2.1   Interdependencies of infrastructure components.

Institutional infrastructure sometimes referred to as social or institutional capital relates to the system of informal and formal rules that govern an organisation or a country. Personal infrastructure refers to 'human capital' – the stock of tacit and explicit knowledge and skills embodied in the workforce, nurtured through investments in education, training, and supported by health and other social services. Physical infrastructure or physical capital comprises all physical elements of buildings, structures and networks – transportation, power supply, sewerage and telecommunication systems, hospital and industrial buildings etc. The understanding of the relationship between these components is crucial in policy making and the delivery of infrastructure services. For example, in the context of delivering healthcare services, providing doctors and nurses, financing systems and the physical infrastructure (hospital buildings) are not sufficient. It is the service derived from the use of such facilities (physical infrastructure) through integrated hard and soft facilities management that is crucial. There is therefore a need for a service-focussed approach in the development of infrastructure projects.

Types of physical infrastructure

Physical infrastructure is classified according to function (see Figure 2.2) or morphological features. Technical infrastructure, often referred to as economic infrastructure, comprises the long-lived networked, capital-intensive and engineered structures indirectly supporting economic production. Trade infrastructure represents the facilities directly used for the production of goods and services such as factories, warehouses, shops and offices. Social infrastructure focuses on facilities directly related to human and social welfare such as schools, colleges, health centres, sports and recreational facilities which are essential in raising living standards, quality of life and human development.

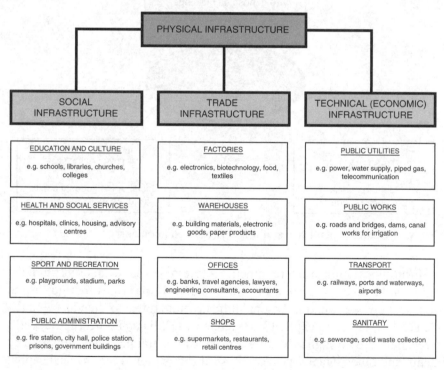

Figure 2.2 Functional classification of physical infrastructure.

There is a need for a systemic approach in understanding and developing infrastructure policies as technical (economic), trade and social infrastructure since these are strongly linked. Other sub-classifications sometimes used include environmental infrastructure (purification plant, refuse incineration plant, rubbish deposit and pollution control and recycling plant), and leisure and recreational infrastructure (stadium, gym hall, swimming pool, zoo, park, museum, theatre, opera and cinema). Administrative infrastructure is sometimes used to classify buildings for government ministries and departments, local government, parliament, city halls, post offices and fire stations.

Physical infrastructure can also be classified morphologically into 'point', 'band and network' or 'space' infrastructure. Point infrastructure is characterised by the location of single facilities in specific areas (e.g. church, mosque, community centre, offices and corner shops). Band and network infrastructure are marked out routes such as roads and railroads, water and sewerage networks, power transmission and distribution networks while space or clustered infrastructure facilities require large areas (e.g. hospitals, university, airport, harbours, shopping complex and business parks). This chapter and the remainder of this book focuses on the nature and development of physical infrastructure as provided by construction and engineering, taking into account environmental, social, economic and political factors.

Infrastructure development

It is expected that the global demand for infrastructure services across all sectors will continue to increase significantly to fulfil household needs and to facilitate economic growth. For example, in roads, railroads, telecommunications, electricity, water and sanitation sectors, Fay and Yepes (2003) estimated that about US$ 370 billion per annum of new investment is needed for 2005–2010, equivalent to approximately 1% of worldwide GDP. About US$ 480 billion per annum (1.2% of global GDP) are also needed for maintenance. These figures do not include other infrastructure sectors, and expenditure on rehabilitation or upgrading of existing infrastructure in the selected sectors. Table 2.1 shows the expected annual investment needs in physical infrastructure by income group from 2005 to 2010.

Table 2.1   Expected annual investment needs 2005–2010

	New		Maintenance		Total	
	US$ Mn	% GDP	US$ Mn	% GDP	US$ Mn	% GDP
Low Income	49 988	3.18%	58 619	3.73%	108 607	6.92%
Middle Income	183 151	2.64%	173 035	2.50%	356 187	5.14%
High Income	135 956	0.42%	247 970	0.76%	383 926	1.18%

Source: Fay and Yepes (2003).

The levels of sectoral investment for infrastructure development also depend on the structure of an economy and the state of development. Countries at the lower echelons of development with predominantly agricultural-based economies tend to have a higher priorities associated with investment in water, sanitation, irrigation, basic health and education projects. As development progresses, the relative share of investment in power, transport, communication, tertiary education, specialist healthcare, sports, entertainment and leisure increases.

There are two basic components to infrastructure development. Provision relates to the strategic planning, regulating and monitoring of the level of services (Ostrom et al., 1993). The level of provision and the types of infrastructure projects selected are influenced by various factors. For example, in education projects the key drivers could be attainment level, literacy rate, retention and absenteeism, and changing employment patterns. In healthcare projects, the drivers could be change of technology, hospitalization rates, level of curative and preventive care, mortality and death rates. These drivers underpinning policy objectives are the key to infrastructure provision, prioritizing infrastructure and identifying actual projects. Production involves transforming infrastructure projects through design and construction, and then maintaining the completed facilities for service delivery.

Understanding infrastructure provision and production processes is crucial in infrastructure delivery. The process requires various inputs such as land, finance, materials, plant resources and professional expertise. A significant amount of time is therefore required to develop provision and production capacity to reduce the risk of infrastructure failure.

Infrastructure development stages

There are various types of project cycles or stages such as those developed by public bodies, professional associations, and international development agencies. However, infrastructure development generally consists of several key stages: (1) planning, (2) design, (3) construction, (4) operational and (5) recycling and disposal. These stages are generally sequential but there are often overlaps between them. The degree of overlap depends on the choice of procurement method. For example, in traditional procurement, there is very little overlap between the design and construction stages compared to design and build or fast track procurement where there are significant overlaps. Implementing infrastructure projects requires input from key agencies in the public and private sectors ranging from project planning and appraisal to operation of the facilities and finally disposal and recycling.

Within each stage there are different actors involved (see Table 2.2) that have various expertise and expectations, which can sometimes be in conflict. The nature or the level of involvement of the various actors depends on the procurement option chosen for the development of the facility.

The need for the involvement of various actors with sometimes conflicting objectives at different development stages/cycles means that the gestation process is often complex and long. Indicative timescales for implementing different types of infrastructure projects are shown in Table 2.3. The operational life of infrastructure facilities also varies considerably from a few years to decades.

Table 2.2   Infrastructure project cycle

Project stages				
Planning	Design	Construction	Operational	Disposal and recycling
Land owner Client (government or private) Users Statutory bodies (e.g. planners, regulators) Developers Design consultants Pressure groups Financiers	Client Architects Engineers Surveyors Statutory bodies (e.g. planners, regulators, fire officers) Facilities manager	Client Architects Engineers Surveyors Statutory bodies (e.g. planners, regulators, fire officers) Project manager Main contractors Specialist contractors Facilities Manager Craftsmen Builders merchants Plant operators	Client User Facilities managers Statutory bodies (e.g. planners, regulators, environmental officers)	Client Statutory bodies (e.g. health and safety, regulators) Restoration specialists Environmentalist

Key participants (row label spanning the table)

Table 2.3   Characteristic times for the various phases of infrastructure development in UK

Infrastructure type	Planning phase (years)	Design phase (years)	Construction phase (years)
Housing	0.5–6.0	0.5–4.0	0.5–4.0
Health	1.0–5.0	0.5–4.0	0.5–5.0
Education	1.0–4.0	0.5–3.0	0.5–2.5
Law courts, civic buildings	1.0–7.0	1.0–3.0	1.5–2.5
Other small and medium buildings (e.g. general offices, telephone exchanges, public libraries)	0.5–3.0	0.5–2.0	0.5–1.5
Roads and harbours	1.5–10.0	1.0–4.0	0.5–3.0
Water and sewerage	1.0–4.0	0.5–3.0	0.5–2.5
Industrial	0.5–2.0	0.5–2.5	0.5–2.0
Commercial	1.0–10.0	1.0–4.0	0.5–3.0

Source: Adapted from Hillebrandt (1988).

Planning stage

The planning stage involves establishing the need and objectives of the project, identifying an appropriate location, seeking outline planning approval for the project, and assessing its technical and financial feasibility.

It is important to justify the need for the project and to have a thorough understanding of the project objectives, the impact of the project on the proposed location, and the investment requirements.

There are three main aspects to consider at this stage – information about the project, information relating to the location of the project and the potential consequences of the project in terms of costs and benefits. Table 2.4 shows some of the key aspects and issues to consider relating to the development of infrastructure projects.

Table 2.4   Key project considerations

Context	Issues
Project information	Client/business needs (utilities consumption, waste disposal, transport etc.)
	Area/land requirements (existing use, future needs, other development etc.)
	Design and construction options (labour, materials, plant)
Project location (environmental view)	Topography and soil (suitability of soil, slope etc.)
	Ground and surface water (quality and availability)
	Climate (wind speed and direction, temperature, flooding, earthquake etc.)
	Pollution (noise, air etc.)
	Plant and animals (biodiversity, conservation requirements etc.)
Project costs and benefits	Capital costs (civil works, plant and machinery, mechanical and electrical etc.)
	Life cycle costs (maintenance, cleaning, disposal, recycling etc.)
	Price benefits (reduce time distance, vehicle depreciation and petrol costs etc.)
	Non-price benefits (increased safety, convenience and pleasure)

The first aspect involves collating and synthesizing information about the proposed project with respect to client or business needs, the space requirements and possible technical solutions. This stage also focuses on identifying possible location for the project and assessing its characteristics in terms of, for example, social, economic, institutional and environmental impact. Table 2.4 provides examples of the environmental factors to be considered. Social and economic factors could include resettlement problems, compulsory purchase of land, compensation, employment and welfare effects. Site selection poses significant problems particularly for large-scale infrastructure projects such as roads, railways, hydropower schemes and airports. In assessing the suitability of the location, it is important to distinguish between the conducive and restrictive location factors as there are different considerations for each type of infrastructure. For example, point infrastructure such as a community centre requires a significantly less area than network or space infrastructure such as roads or university campuses. Table 2.5 is an illustration of

Table 2.5   Examples of conducive and restrictive location factors

	Conducive factors	Restrictive factors
Housing infrastructure	Good ventilation/air movement	Good soil fertility
	Flat to mild slope	Vulnerable areas (earthquake, flooding)
	Good soil bearing capacity	Rich in flora and fauna
Irrigation infrastructure	Fertile soil	Rock/sandy soil
	Adequate rainfall	Contaminated surface and ground water
	Large parcels of land	Too steep

conducive and restrictive location factors for housing and irrigation infrastructure projects.

The final aspect involves assessing the project's viability. It is crucial to establish the different types of costs as well as benefits associated with infrastructure development (see Table 2.6 for an example).

Table 2.6   Easy-to-price/difficult-to-price costs and benefits matrix

Costs	Benefits
Easy-to-price costs	*Easy-to-price benefits*
e.g. Planning	e.g. Increased revenue/savings
Engineering/Architectural design	Rent, income, charges
Construction	Increased land value
Difficult-to-price costs	*Difficult-to-price benefits*
e.g. Air pollution	e.g. Increased safety
Scenic values lost	Increased pleasure/comfort
Accidents caused	Human lives saved

Evaluation of infrastructure projects in terms of the easy-to-price/difficult-to-price cost–benefit matrix at this stage will ensure that development is both feasible and sustainable based on assessments reflecting the implications of preliminary design variables in terms of structural, geotechnical, environmental, and social considerations. For example, the initial cost of power generation projects includes the construction cost of generation plant, transmission and distribution systems. The whole-life-cycle or total cost of infrastructure development generally involves the following:

- Land costs (land price, associated legal fees, compulsory purchase)
- Planning costs (planning application fees, planning consultant fees, building regulation fees)

- Design costs (architects, engineers, surveyors, management fees etc.)
- Construction costs (materials, labour, plant, management)
- Operating costs (running costs, maintenance costs)
- Recyling and disposal costs.

Establishing the potential for cost recovery/revenue streams and profit in relation to the risk is crucial in determining the financial viability of infrastructure projects. The project planning stage has been identified as the main cause of lower economic returns in infrastructure projects. In the case of complex infrastructure projects it is important to arrive at a workable project finance structure striking the right balance between equity, debt, public and private finance. This is a complicated process and will require the expertise of specialist financial advisors.

Poorly defined and planned infrastructure projects often have unclear aims and objectives that can create difficulties in the short and long term, resulting in potentially irreversible damage and consequences. On the other hand, sensible decisions made at this stage concerning such matters as project definition, location and investment requirements could have long-lasting positive effects on infrastructure development.

Planning permission is required for virtually every development. This will involve extensive consultation and interaction between clients, their advisers with relevant planning and regulatory authorities. It requires an understanding of appropriate legislation and policies, knowledge of how planning processes operate and the planning framework, which usually consist of several elements or layers. In many developed countries it is usual for some kind of structure plan to provide general policies or strategic needs relating to developments over a period while local plans reflect strategic needs by allocating specific sites for development. Outline planning applications for infrastructure projects are assessed against both structure and local plans which should be regularly monitored and reviewed.

Design stage

The design stage involves developing solutions to reflect the client's brief and the planning parameters and constraints outlined in the planning stage. This follows a certain sequence and iterative process from the evolution of the design concept into a scheme design from which a detailed design solution can be developed showing proposed physical features, functional characteristics, and incorporating specialist design options. Expert design input is provided by architects, engineers, surveyors and planners but increasingly important is the input from facilities management consultants. A detailed cost plan is also provided and an assessment of the construction processes and techniques required for project implementation is usually undertaken. Detailed design decisions are made or fine-tuned based on the shape and dimensions of the facilities, aesthetics, the regulations the facility should comply with, how the facility should look in terms of the quality of materials and components to be used and how it will be constructed.

This stage also involves submitting detailed planning permission to the planning authorities based on the detailed designs or modifications to ensure compliance with the requirements of various legislation such as building regulations, health and safety. Once the drawings are completed, the contract documentation process puts together the design drawings, detailed specifications, bills of quantities, schedule of rates, general and specific contract conditions. These documents usually form the basis for inviting bidders but the activities that take place during bidding and the type of documents depends on the procurement choice, whether it is open competition, limited competition, or some form of short-listing using a pre-qualification process or negotiation.

Construction stage

The construction stage involves managing various resource inputs such as labour, materials, components and plant to produce the end-product. The level and type of inputs, together with the implications for co-ordination and management structure depends on the nature of the project or the end-product. For example, building type infrastructure tend to require greater co-ordination from a range of craftsmen (carpenters and joiners, mason, electricians, plumbers, glaziers, painters roofers, scaffolding specialists) compared to non-building type infrastructure such as roads and railways with limited range of craftsmen. There is a range of 'end products' from small and simple buildings to large and sophisticated structures such as bridges and dams. Construction 'end products' may be classified into three distinct types: standard construction; traditional construction; and innovative construction (Bennett, 1991).

There are different technical and management processes required in infrastructure construction depending on the end product. Technical processes range from approaches relying mainly on tacit knowledge of individuals to automated processes relying on intelligent and knowledge-based systems such as robots for on-site construction. Management processes depend on the type of construction. Standard construction projects require programmed organisations relying heavily on routine and standard procedures to manage the construction process, while innovative construction requires highly flexible matrix management procedures to manage complex processes. The final activity is commissioning and handing over to the client, which in turn is linked to the operation of the completed infrastructure facility. This involves testing of the systems in the facility, providing appropriate operational manuals such as a safety procedure manual, training of personnel, removal of plant and equipment, temporary works and storage, and to ensure the completion of any other external works.

Operational stage

The delivery of the commissioned project should be the full realisation of the client's aims and objectives as represented by the project brief. The next stage is to operationalise the project according to the relevant performance benchmarks, operational targets and the expectation of users of the facility. The responsibility for the operation of the infrastructure facility will depend on the procurement method selected. Where all responsibility is handed over to the client at the end of the construction stage, with the exception of defects liability, then the client will be required to arrange for total facility management. In the event that the contract covers design, build, finance and operate (DBFO) or a derivation thereof, then the contractor (concessionaire) will be responsible for the operation of the infrastructure facility according to a service level agreement that extends over an agreed concessionary period. Whoever is responsible, it will be necessary to take into account changes in business activities, space requirements, increased user flexibility, technological development and the fact that the end-users of facilities are often different from the owners makes it an extremely crucial stage. This stage places emphasis on adequate facilities management and the extent that this expertise has been taken into account at the design stage.

Given that construction of an infrastructure facility is not an end itself, it is now increasingly recognised that the services derived at the operational stage need to be sustainable over the whole life cycle of the project and will need to address issues relating to the conservation of non-replaceable resources. Typical activities at this stage include collection of user fees, repairs and maintenance, cleaning, soft and hard landscaping, security and support facilities including catering, housekeeping, and car parking.

Recycling and disposal stage

The disposal and recycling stage is when the project has reached its intended design life. At this stage, consideration should be given to how facilities could be reused, how any component, material or part of the facility could be recycled and the associated cost. Recycling the facility may involve some alterations to update the facility, which may require further planning approval. This stage will involve the identification and recycling of recovered products and materials for use in other facilities, which may be for both high and low-grade applications. The present trend in the recycling of construction materials is likely to continue given widespread interest in sustainability issues. There is also a need to identify hazardous or potentially explosive processes, noise and fume problems in disposal and compliance with regulations. This stage is central to the sustainability debate. Sustainability has become a major issue in infrastructure development as it is increasingly recognised that environmental and social considerations should not only be integrated into design and construction processes but it makes good business sense.

Toward a service-focused approach

In the traditional procurement approach, the responsibility of construction companies terminates when design and construction works are completed and obligations under defect liability are fulfilled. The facilities are then transferred to the owners often without a detailed knowledge of how best to integrate soft and hard facilities management to obtain the best in use performance.

While the traditional approach continues to be the most popular procurement method, given the right circumstances, many major construction companies have recognised opportunities to provide a holistic procurement. This approach incorporates design, financing, construction and post construction asset services or facilities management over a predetermined concessionary period in exchange for rent or the benefit of user charges, or a combination of both. Such concessions typically cover 10–30 years and may be extended if both parties agree. Hence, greater emphasis is placed on operational performance and efficiency that has engendered a shift to a 'service focused' approach.

This 'service oriented' approach is crucial particularly in the light of alternative procurement methods such as 'build operate and transfer' (BOT), 'build, own, operate and transfer' (BOOT), and more recently public–private partnerships (PPP), an example of which being 'private finance initiatives' (PFI) in the UK. Typically, PPP projects are based on the concept of continuous provision of adequate level of services to render the facilities operational at all times and to make the investment worthwhile. These new approaches transfer the responsibility for asset management, replacement, renewal or disposal from the public to the private sector often with better incentives and management structures.

These changes reflect the need for a greater integration of the project stages during infrastructure development. The implications are that there is now a need for greater integration of planning, design and construction, operational, recycling and disposal stages and to view infrastructure as a system to facilitate the delivery of services rather than as an end-product. Infrastructure is about systems as each component (physical, institutional and personal infrastructure) cannot operate independently – the interrelationships between physical aspects of the environment (physical infrastructure – buildings and networks), rules and regulations (institutional infrastructure) governing the provision of particular service and the human inputs (personal infrastructure) are crucial in a service-focussed approach.

A service-focussed approach means that there is also a need to shift performance measurement from the level of provision (e.g. number of hospital beds per population) to more appropriate end-user service performance indicators (e.g. availability/functioning of operating theatres). These indicators, which could be developed for various types of infrastructure should be measurable to reflect the characteristics of the service in terms of quantity, quality, reliability and coverage. The continuing growth of PFI procurement in the UK and other types of service contracts in developed and developing countries means that some of these service level indicators already exist, while others are gradually evolving and being tested in key sectors.

Case studies

Case study 2.1: Mexican toll road programme

Mexico's private toll road programme more than doubled the national toll roll network from 4500 km in 1989 to 9900 km. Fifty-three concessions were awarded for approximately 5500 km of roads and by 1995, 45 were in partial or full operation representing 5120 km. The total investment for the period was US$13 billion obtained from domestic commercial banks, concessionaire equity, federal and state government grants and equity. The macroeconomic project level factors meant that new development of another 6500 km needed by 2000 could not progress as a result of severe financial and economic consequences, huge non-performing loans from previous concessions owed to commercial banks estimated between US$ 4.5 to US$ 5.5 billion.

The main problems were as follows:

- The toll road programme was not adequately planned due to a rush to develop infrastructure quickly to compete effectively in the regional free trade zone.
- The programme was used as an instrument to lift the construction industry out of economic depression.
- There was a significant shift of capital flows to the road infrastructure sector which increased the commercial banks loan portfolios.

There were a number of factors contributing to the poor planning, design and implementation of the road programme (Ruster, 1997). There was a serious lack of intermodal development policy and strategy integrating rail, port and airport transport as a result of inadequate planning criteria at the federal and state level. There was very limited preliminary engineering, unsatisfactory review of traffic studies, geotechnical and environmental studies, and the tendering process and concession design was inadequate. The pre-qualification process was vague, and at times opaque, making it extremely difficult to screen out bidders that lacked the capacity to manage risks of design, construction, management and operation of toll roads. There were also problems relating to uncertain tariff adjustment

procedures, and a lack of independent regulatory authority to supervise contractual arrangements.

Inadequate planning also led to poor design solutions, contract
documentation and procurement which were exacerbated by the
problems identified above. The project award criteria limited competition to a handful of major construction organisations that were only
interested in the construction aspect without a full evaluation of their
capacity to deal with the operational aspects and long-term viability
of the programme. Inadequate consultation created resistance from
the local community, including farmers, environmentalists and other
special interest groups during construction which prompted the
re-routing of some road schemes due to difficulties associated with
securing rights of way. Understaffing and limited institutional capability within the Ministry of Transport meant that there were inefficient contract management procedures and processes typified by
information deficiencies, difficulties in managing design changes
unilaterally mandated by Ministry, delay in approving variations,
inadequate enforcement regimes regarding the quality standards
for construction and maintenance. Lack of technical capabilities in
terms of specialised machinery, skilled labour, quality control procedures created a situation were the handful of large construction
firms involved in the road programme were stretched to their limit as
a result of the speed at which concessions were being awarded. An
underdeveloped institutional capability and the programme scope
and content that exceeded the technical and administrative capacity
of the local construction industry meant that there were excessive
cost overruns and delays. As many projects were tendered on a
cost-plus basis, the average cost per kilometre of road rose sharply
by over 50% compared to the original estimate.

Although the programme initially attracted significant private sector
interest and investment, the well publicised problems significantly
affected the performance of the road programme. The relationship
between private and public sectors were not transparent and at
times adversarial. Investors were unable to determine or understand
how projects could fit into long-term development plans. Liquidity
problems associated with domestic financial markets limited the
project finance experience of most financial intermediaries involved.
Furthermore, limited institutional capabilities of regulatory officials
exacerbated the funding problems. Inadequate financial discipline
in government-owned commercial banks meant that important
conditions critical to funding were waived (e.g. insurance, bonding
requirements, unsatisfactory review of traffic studies, geotechnical
and environmental studies) as there was an attitude that government will always absorb risks even if projects become commercially
non-viable. The situation was made worst by the currency crisis in

1994 when all-in interest rates rose to over 100% a year for most projects leading to severe debt service problems. Gross miscalculation of investment costs and operating income resulted in an unsustainable set of operating conditions. The implication was that concessionaires were faced with writing off significant portions of their investment and the government was under severe pressure to rescue investors by injecting financial resources. The outcome resulted in users paying some of the highest toll charges in the world.

Author's commentary
The widespread difficulties experienced in the implementation of the Mexico toll road projects suggest that there was a need for an appropriate infrastructure policy framework to facilitate the implementation of projects. The absence of a robust policy framework and a strategy meant that there was no basis to facilitate the participation of the private sector in planning, design, construction and the operation of toll road projects. As a result, the implementation of the road programme was flawed from the planning to the operational stages. Such failures were characterised by lack of adequate consultation, stakeholder participation, poor co-ordination between different agencies of governments involved in transport, incomplete design, poorly developed contract and procurement documentation, inadequate resources in terms of qualified manpower to monitor the development and implementation of the toll road projects. There was also a systemic failure to recognise that the most important aspect in defining the objectives of the toll road projects was the service delivery at the operational stages. Instead there was a disproportionate emphasis on the construction aspect with very limited attention paid to the experience and capability of bidders on the operation and maintenance of toll roads i.e. service delivery.

Nigeria too experienced similar problems as a result of having to implement hastily planned infrastructure projects, without adequate planning, regulatory and construction capacity during the oil boom. The result was costly both in terms of delay in implementing projects and the utilisation of investment. Previous examples in the mid-60s include the experience of Singapore in implementing a major infrastructure development plan for roads, bridges, drainage and housing which was not fully implemented due to local construction resource capacity.

Case study 2.2: Oslopakke 2: Strategy for better public transport in Oslo and Akershus, Norway

Oslo and Akershus County has a population of 1 million inhabitants and although the land area represents only a small proportion of

Norway it accounts for 20% of the population. Progressive migration of people to this region over the past 25 years has placed considerable demands on old and worn out infrastructure. This has resulted in inadequate capacity, congestion and poor punctuality. Furthermore, serious environmental problems have been created and there is a recognition that road capacity expansion schemes alone cannot solve these problems. Therefore the challenge is to improve public transport sufficiently to reduce road use and thereby improve air quality and noise, especially in central areas.

In order to encourage increased use of public transport the following improvement in standards need to be addressed:

- increased speed, safety, regularity and functionality;
- more seats and better comfort;
- higher frequency of the main network;
- public transport for new residential and workplace areas;
- better co-ordinated timetables for buses and trains; and
- improved transfer between different modes of public transport and more attractive stations and bus stops.

In 1990 the Oslo ring road (Oslopakke 1) was established to assist the development of the main road system in the region and to donate 20% of its revenue to improve public transport, especially buses, trams and the Metro.

National and local government introduced Oslopakke 2 in 2001 with a view to providing funding through user surcharges to upgrade infrastructure and rolling stock for public transport.

The financial framework provided by Oslopakke 2 between 2002 and 2011 will amount to 11.3 billion Norwegian Krona (NOK) and will be sourced as follows:

71% National rail and road infrastructure budget
5% Oslo's City Budget
3% Property developers
21% Payments from road and transport users

Of the total investment 60% will be on improvements to the rail track system, stations and terminals. New rolling stock will account for 8% and approximately 32% will be spent on local infrastructure such as the Metro, multimode terminals and traffic flow measures, including bus lanes and signal control.

The main projects will be:

1. West Corridor Railway at a cost of 6.5 billion NOK (2002). There will be 19.5 km of parallel existing and new tracks running through 14.5 km of tunnel. Four stations will be connected and construction work is due to be completed in 2011.

2. Extending the existing line from Ulleval Stadium to Carl Berners will provide a new Metro line. This will complete the Oslo circle line. 5 km of line (4 km in tunnel) will be undertaken in two stages at an approximate cost of 1 billion NOK.

A ticket surcharge on public transport will be made to provide funding for the purchase of new rolling stock and increased seat capacity.

Author's commentary

Oslopakke 1 and 2 demonstrate national and local government awareness of the need to accommodate economic, demographic, social and environmental change by making investment in physical infrastructure. Problems concerning the provision and performance of existing transport infrastructure have been anticipated and identified to enable proper policy and strategy to be developed for renewal, expansion and maintenance to accommodate increasing demand. The key objective has been to make public transport more reliable, affordable and safe, while reducing adverse affects on the environment caused by pollution from traffic congestion.

A combination of public and user finance has been adopted to provide sufficient resources to enable aims and aspirations to be achieved. This case study emphasises the need for strategic planning in accordance with a holistic understanding of the environment, change and the needs of both the nation and local communities.

Summary

This chapter has discussed the concept of infrastructure, emphasising the need for a service delivery approach. The implication for policy making is that there is now a need to view infrastructure as a system to facilitate the delivery of services rather than as an end-product. A classification of physical infrastructure based on functional and morphological features has been presented. The key stages of infrastructure development, associated issues and the actors involved from planning to recycling and disposal, as well as their interdependence in ensuring a successful implementation of projects was also discussed. The chapter concluded with case studies reflecting the need for a policy and strategic framework to enable wider considerations beyond the design and construction stages for the development, implementation and sustainability of infrastructure projects.

References

Bennett, J. (1991), '*International construction project management: general theory and practice*', Butterworth-Heinemann Ltd, Oxford.

Fay, M. and Yepes, T. (2003), '*Investing in infrastructure: what is needed from 2000 to 2010*', World Bank Policy Research Working Paper 3102, July, World Bank, Washington, D.C.

Hillebrandt, P. (1988), '*Analysis of the british construction industry*', Macmillan Press, London.

Jochimsen, R. (1966), '*Theorie der Infrastructur, Grundlagen der Marktwirtschaftlichen Entwicklung*', Tubingen, p 145. Citcd in UNCHS.

(Habitat) (1989), '*Methods for the allocation of investments in infrastructure within integrated development planning: an overview for development planners and administrators*', Habitat, Nairobi.

Miller, J.B. (2000), '*Principles and practice of public and private infrastructure delivery*', Academic Publications 2000, Kluwer.

Ostrom, E., Schroder, L. and Wynne, S. (1993), '*Institutional incentives and sustainable development: infrastructure policies in perspectives*,' Boulder-Westview Press.

Ruster, J. (1997), '*A retrospective on the Mexican toll road program (1989–94)*' Viewpoint, Note No. 125, September, World Bank.

Sloman, (1991), '*Economics*', Hertfordshire, Harvester Wheatsheaf, Prentice-Hall.

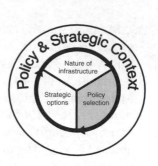

Chapter 3

Determination and selection of infrastructure policy

Introduction

A key issue in the delivery of infrastructure is the capacity to develop and implement policies as it has a significant influence on the level of co-ordination, financial and technical resources available to improve the level of services. To enable countries to develop infrastructure at a national and local level, an effective policy framework is necessary to identify specific infrastructure needs, facilitate the selection and implementation of infrastructure projects, and to monitor the performance of infrastructure assets and services. Developed nations have been able to strengthen their policy capacity where there is a need to improve infrastructure delivery, to remain internationally competitive and to cope with the increasing challenges of globalisation. However, developing nations have had to rely increasingly on the policy expertise of development agencies such as the World Bank, regional development banks, multi-lateral and bi-lateral development agencies to improve their policy capacity. This is vital in the context of globalisation to facilitate integration with developed nations, and to increase both private sector and foreign direct investment necessary for economic growth and the reduction of poverty. For example, the Rapid Response Unit recently set up by the World Bank is in recognition of the need to improve policy capacity in developing countries in terms of the design and implementation of infrastructure projects. This unique knowledge resource provides access to a global community of policy makers and advisors, and acts as a forum for debating critical policy issues.

Establishing a policy framework is therefore fundamental to improving the efficiency and effectiveness of infrastructure delivery. This is crucial in establishing the needs, improving the level of services and assessing the impact on all stakeholders involved. Central to the policy making process are the institutions, goals and objectives, resources, knowledge, information and communication systems. The environment is also crucial as it reflects the economic, social, and political context that influences the policy-making process. The formulation and implementation of appropriate infrastructure policies therefore requires an understanding of these key elements and their interactions. Following this introduction, the policy context for the delivery of infrastructure services is examined in terms of the environment, goals and objectives, information and communication systems, and resources. The factors necessary to develop or strengthen policy capacity, to improve its effectiveness in order to continuously attract investment and to deliver sustainable infrastructure services are explored. Also discussed are the key criteria that determine public or private participation in the delivery of infrastructure services. It is argued that decisions on whether the private or public sector should provide a particular type of infrastructure are rarely based on economic considerations alone but a range of factors such as public service obligations, externalities (third-party costs and benefits), national security and defence.

The policy framework

One of the most fundamental issues in the delivery of infrastructure concerns decisions on what types of infrastructure are required and how they should be provided. The policy framework influences the level of infrastructure provision and production and depends on policy objectives, the implementing institutions, level and type of resources, knowledge, information and communication systems, and the environment.

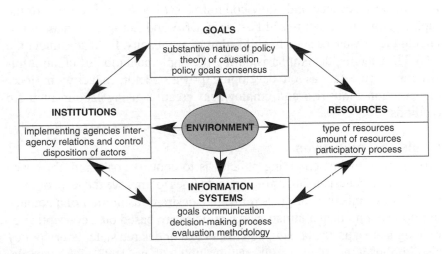

Figure 3.1   Elements of the policy-making process.

Environment

The environment is made up of individuals, groups and organisations often with different views, values and sometimes conflicting interests, operating as a system. Understanding the environmental aspects surrounding the

33

policy-making process and its impact is the most crucial factor in developing and implementing policy goals and objectives. There are different stakeholders with a range of views such as governments, private investors, regulators, design and construction firms, consumers and users, and special interest groups like trade unions, environmentalists and conservationists. Depending on the issues different stakeholder interest may impact on the success or failure of a policy. For example, a policy aimed at attracting significant investment for infrastructure development may give higher priorities to the views of the private sector. This is because the private sector tends to be more attractive to investors as they are often considered to be better at risk management, more commercially aware and more efficient than the public sector. Political factors such as government commitment are also central to the environment, so are the socio-economic considerations such as level of employment, poverty, regional distribution and equity objectives. There are other internal and external factors that impact on the environment such as international trade agreements and other international protocols.

The environment also influences the level of private sector participation and the motivation for local and foreign investment. Infrastructure investment patterns differ in Advanced Industrialised Countries (AICs), Newly Industrialised Countries (NICs) and Least Developed Countries (LDCs) as private participation in infrastructure is sometimes promoted for a variety of reasons that in turn may be influenced by the environment. In Latin America and Eastern Europe, the main driving force was mainly due to the need to raise revenue and show commitment to economic reform. In the rapidly growing economies of East Asia, USA and Europe, the main influencing factor was the need to build additional capacity. In Africa, under the AGETIP scheme, the emphasis during the implementation of urban infrastructure projects was on developing private sector capacity for local construction industries and employment creation using labour-intensive methods.

Goals and objectives

The essence of implementing policies is to convert goals and objectives into real and beneficial infrastructure outcomes to improve the level of services. Policy objectives or goals reflect the desire or aspiration of a country to improve a particular situation. All policies are based on a conception of moving from a particular situation to an ideal or desired state. Every policy also implies a theory or causal relationship (Jenkins, 1993). For example, a policy for creating employment based on expertise and skills must be underpinned by an adequate theory of employment creation and capacity building, and cause–effect relationship between the policy variables and employment outcomes. However, if the theory or cause–effect links are incorrect, the policy will fail irrespective of how well it is implemented. It is therefore important to have a thorough understanding of the problem, the theory underlying the problem and its context, in order to be able to formulate robust policy goals and objectives.

Sectoral assessments and analysis of the state of infrastructure is required to determine gaps and to provide the basis for understanding infrastructure development problems in order that appropriate development goals and objectives can be formulated. There may be conflicts between ideologies, political groups, different agencies and special interest groups in the formulation of such objectives. These differences must be recognised and addressed. In the end, the goals and objectives arrived at should be as a result of the interactions and consensus between various stakeholders in the development process such as civil servants, politicians, planners, investors, lenders, engineers, lawyers, regulatory specialists, environmentalists etc. The policy goals and objectives are often reflected in various types of documents or instruments such as policy statements, structure plans, master plans, regional plans to implement infrastructure programmes.

Knowledge, information and communication systems

Knowledge and information relating to policy goals and objectives should be clear and readily available for communication to providers, users and stakeholders. The level of knowledge and the quality of information will have a significant influence on the nature and outcome of policies, together with their subsequent monitoring, evaluation and feedback. Information is central to making informed choices about the effects of policies. At the micro level, it may be necessary during the exploration of policy options to collect information on willingness-to-pay for particular infrastructure services such as transport, patterns of consumption of water, electricity and gas. There may also be a need for socio-economic data e.g. educational attainment, income, employment and family sizes. Other information may be relevant, for example country risk assessment profile, market information on specific sectors, database of infrastructure projects and land development rights for network infrastructure e.g. roads, railways. The credibility of such data may influence foreign or private investment. It is therefore important to be aware of the risk of information being sometimes unreliable, inadequate, non-existent, or deliberately withheld, particularly in some developing countries. The lack of adequate information on labour-based methods to make cost and time comparisons of alternative construction methods has often been cited as a key factor affecting the decisions of engineers responsible for design and construction, especially in developing countries. The information component is therefore crucial in developing and evaluating policy alternatives, establishing trade-offs and making knowledgeable decisions about policy choices.

Communication systems that include IT and non-IT systems are also essential to regularly inform stakeholders about policy updates and changes. Non-IT systems include policy briefing sessions, seminars, newsletters and press releases. IT systems are useful for storing, processing, assessing, monitoring and evaluating information concerning infrastructure projects and their implementation. Communication systems facilitate access to information that is vital to the reduction of transaction costs and the creation of competition in infrastructure projects. Communication systems based on information technology have an increasingly important role to play in infrastructure delivery. Global trends in trading goods and services, including that of infrastructure investment, means there is vital need for adequate communication systems worldwide. The internet has a key role to play since it can be used as a repository for storing information on all aspects of infrastructure policies and projects. The new internet portal (IWorld Infrastructure World.ComInc) co-created by the International Finance Cooperation (IFC), reflects a major change in global infrastructure markets by linking buyers and sellers of infrastructure development around the world. This global portal provides an electronic platform for business transactions throughout the life of infrastructure projects and involves the active participation of developing countries that need infrastructure, and multinational or transnational companies and investors in developed economies that have the funds to build and provide essential infrastructure services. The portal also contains specialised websites for advice on infrastructure project development and an on-line directory of consultants and advisors.

Resources

Physical infrastructure projects require a wide range of resources covering materials, components and equipment to human labour, expertise and skills, together with the necessary funding to undertake the work. It is therefore essential to determine that sufficient resources will be available to implement policy objectives.

Specifically, there is the need for a range of human resources for infrastructure provision relating to planning and organisation (e.g. planners), arranging finance (e.g. legal and financial specialists), regulating and monitoring of infrastructure services (e.g. safety, environmental and regulatory specialists). The resource implications of an infrastructure investment programme should also be assessed in terms of capacity for production such as the technical skills required (planners, architects, engineers, quantity surveyors, construction crafts, construction managers, facilities managers, etc.).

Physical infrastructure projects will require an adequate supply of construction materials, components and equipment. When these are not locally available they will need to be brought in from other regions or imported. The extent to which new technologies are to be used will depend on the design solution required to provide infrastructure that is feasible and affordable. Consequently, it will be necessary to identify individual suppliers who collectively will form a holistic supply chain. This is particularly necessary in transition and developing economies where resource markets are often underdeveloped and unpredictable leading to significant increases in infrastructure development costs. Morah (1996) highlighted problems related to resources as the main factor for policy implementation gaps in developing countries where resources are scarcer.

The resource requirement or availability is strongly related to institutional options selected for the delivery of infrastructure services. Different institutional arrangements have different implications for resources in terms of personnel, capital investment, recurrent expenditure for the maintenance and operation of infrastructure facilities. For example, under the purely private enterprise option, assets are owned by the private sector usually through new investment financed by BOO or BOT schemes, or the privatisation of existing public assets. The responsibility for further capital investment, recurrent investment, provision of technical and managerial personnel and associated risks management expertise lies with the private entrepreneur. There are various options relating to public, private and public–private partnerships to attract investment, increase efficiency and to expand infrastructure services, depending on the policy framework. Table 3.1 provides a more detailed comparison of the resource implications of different institutional forms. Implications for planning and executing investments, capital financing, operating and maintenance cost are in rows (2) to (6). Co-ordinated public and private investment plans may be established detailing the resource implications of various types of infrastructure projects, and the likely or realistic implementation periods based on the institutional options selected.

Institutions

Institutions define the implementing agencies, whether public, private or mixed, the locus of responsibility, the nature of inter-agency relationships and the actors. Inter-agency relations are crucial, so are the number of actors and institutions involved. Poor co-ordination between government agencies both at central and local government and the private sector can sometimes delay the implementation of infrastructure policies and projects. Too many institutions with diverse perspectives and interests serve to increase the problems of co-ordination, which could muddle policy objectives. There is increasing evidence of major institutional changes concerning infrastructure development with a view to continuously improve the level of services to support household needs, economic growth and productivity. Table 3.1 shows a range of institutional options with different implications for ownership of assets, managerial authority, policy making and risk management. However, from an institutional perspective, it is important to distinguish between provision and production of infrastructure. Provision relates to the planning, financing and monitoring and regulation of infrastructure (Ostrom et al., 1993). A publicly funded infrastructure facility requires funding from the government's budget even though it may be *produced* by the private sector. Production on the other hand focuses on the design, construction, maintenance and operation of the facilities for service delivery. Public goods such as street lights, traffic control systems are

Table 3.1 Comparison of institutional forms by assignment of functional responsibility and resource implications

FUNCTIONS	Government department	Parastatal/public utility		INSTITUTIONAL FORMS — Service contracting	Management contracting	Leasing	Concession	Co-operative/ commercial	Private entrepreneur
		Traditional	Corporatised commercial						
1. Ownership of assets	State	State	State (majority)	State or mixed				Private or in common	Private (majority)
2. Sectoral investment planning and policy making regulation	Internal to government	Parent agency		Parent agency or separate public authority			Public authority negotiated with contractor	None or public authority	None or public authority
3. Capital financing (fixed assets)	Government	Subsidies/ government	Parastatal limited subsidies	Public	Public	Public	Private contractor	Private	Private
4. Current financing (working capital)	Government	Backed borrowing	Market-based financing	Public	Public	Private contractor	Private contractor	Private	Private
5. Execution of investment	Government	Parastatal	Parastatal	Private contractor for specific services	Public partner	Public partner	Private contractor	Private	Private
6. Operation and maintenance	Government				Private contractor	Private contractor	Private contractor		
7. Managerial authority	Government	Government	Parastatal	Public partner	Private contractor	Private contractor	Private contractor	Private	Private
8. Commercial risk	Government	Government	Parastatal	Public partner	Mainly public	Private contractor	Private contractor	Private	Private
9. Duration	No limit	No limit	No limit	<5 years	3–5 years	5–10 years	10–30 years	No limit	No limit

Source: Kessides (1993).

produced by private firms and sold to the public sector. Roads, schools, hospitals and national parks provided by the public sector may also be produced i.e. designed, constructed and maintained by the private sector. The share of public/private provision and public/private production varies significantly from country to country. In advanced Western economies the tendency has been to move production from the public to the private sector, primarily through PPP type initiatives, whereas in the former communist or socialist economies the bulk of production remains in the public sector. Nevertheless, this situation is changing and the trend is to involve the private sector in the provision and production of public infrastructure at every opportunity.

The contrasting scenarios shown in Figure 3.2 will result in different levels of interactions between actors in the public and private sectors.

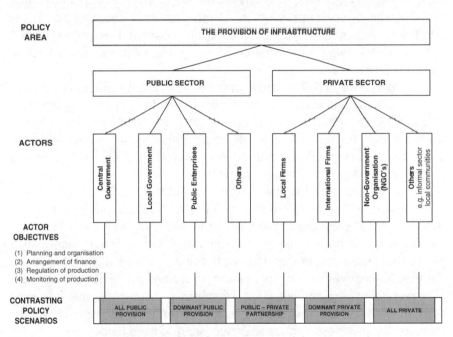

Figure 3.2   Policy framework for infrastructure provision.

The policy framework should be robust to support the different institutional options available to improve infrastructure delivery (Miller, 2000). Public–Private Partnerships such as the private finance initiative (PFI) in the United Kingdom have encouraged innovative relationships between public and private enterprises to facilitate significant investment from the private sector in the provision of schools, hospitals, transport, and other sectors (Broadbent and Laughlin, 2003; Grout, 1997). Private Participation in Infrastructure (PPI) in developing and transition economies, which initially focused on power and telecommunications subsectors has now been successfully extended to other subsectors. Private investment has increased significantly over the past few years, and the trend is expected to continue. The provision of economic (technical) and social infrastructure services is still dominated by the public sector but trade infrastructure has become the responsibility of the private sector in market-based economies. Following the collapse of the communist regimes in Eastern Europe, significant trade infrastructure (oil, gas, steel and others) have been sold to the private sector.

Determinants of public and private participation

Water, sewerage, roads, education, health, recreation, electricity, industrial and other infrastructure services are all defined as goods according to classical economic theory. For historical reasons, economic and social infrastructure such as highways, sewers, hospitals are usually referred to as public capital (goods), whereas trade infrastructure – factories, warehouses, private offices, retail and wholesale shopping facilities are considered as private capital (goods). But distinguishing between what is public or private infrastructure capital is problematic.

A central issue in policy development is the determination of the respective roles of the public and private sectors in the delivery of infrastructure. The late 1990s were characterised by a significant shift from the public to private sector for the provision of certain infrastructure services due to a number of reasons (World Bank, 1994; Kessides, 2004). First, there was consistently poor performance in public enterprises due to the lack of expertise in commercial and risk management. Second, there were significant constraints, leading to severe shortfall in public sector budget for capital investment and recurrent expenditure to maintain infrastructure facilities; the result was a huge backlog of maintenance in key infrastructure such as transport networks, power stations, hospitals, schools. Third, there were increased incentives for the private sector to assume and better manage risk. However, there are a number of key economic and non-economic factors to consider in determining whether a particular infrastructure should be provided by the private sector. These factors are discussed below.

Principles of excludability and subtractability

The principles of excludability and subtractability are economic tools that are used when making decisions on whether goods or services are to be provided publicly or privately. The benefits generated by an infrastructure facility are said to be 'excludable' if people can be easily prevented or restricted from using the services. The non-excludability of the benefits generated from particular goods is often cited by economists as the hallmark of a good that must be provided publicly. Non-excludability means that a user cannot be prevented from using or benefiting from a facility because it is either extremely difficult or expensive to exclude people. This is the so-called 'free rider' problem and classic examples of such goods are street lighting, national parks, radio broadcasting and communication systems. While it is impossible to exclude people from benefiting from street lighting, it may be possible to exclude people (although at considerable cost) from visiting national parks due to certain features and the enormous size of the boundaries. However, in the case of radio broadcasting the requirement to have a radio receiver (often referred to as access or intermediate goods) restricts the benefits from such public goods.

In theory, most infrastructure facilities are in a sense excludable, as their use depends on gaining access to a facility or network, which can arguably be restricted. The degree, practicality, ease or the cost of excluding people depends on the morphological characteristics and technical features of the facility. For example, it is much easier to exclude people from gaining access to point infrastructure such as swimming pools compared to space infrastructure such as university campuses or national parks. Advances in technology such as the development of sensors, intelligent systems and hi-tech barriers means that it is now possible to exclude people (at reasonable cost) from gaining access to large spaced facilities such as University campuses. It would have been extremely difficult or expensive to exclude drivers from using urban roads in the past, but recent advances in technology, particularly in the areas of sensors have made it easier to implement exclusions at reasonable cost. The recent introduction of congestion charges in central parts of London and Singapore are examples of the application of the exclusion principle. Although network infrastructure such as electric power and water supply systems are large and provide access for the population as a whole, their special technical features makes it extremely difficult to benefit from such facilities without paying for them.

Subtractability in theory means there is rivalry in the consumption of infrastructure services. Non-subtractability means that consumption by one user does not reduce or limit consumption of that same good by others. A classic example is the consumption of street lighting by one individual which does not affect in any way the amount of lighting available to other road users or pedestrians. Conversely, the consumption of water from an irrigation canal by one farmer means that there is less water available for other farmers to use. In the case of urban roads, as more drivers continue to use the same road or section of that road then congestion will occur thereby reducing the benefits to other road users i.e. the presence of other drivers affects the journey time of an additional driver.

Pure public goods are defined as those that are both non-excludable and non-subtractable, whereas a purely private good has neither characteristic. The benefits derived from public goods are not limited to the purchasing individuals, as is the case with private goods. This may be illustrated by improvement in transportation infrastructure intended to reduce air pollution that will benefit many people but the consumption of good-quality air by one individual does not reduce the benefit available to others. In the case of pure private goods such benefits are 'internalised' and therefore not available to others.

		Degree of Subtractability/Rivalry	
		Low	High
Degree of Excludability	Low	**PUBLIC GOODS** Road signaling equipment Street lighting National parks Radio broadcasting/communication system	**COMMON PROPERTY** Urban roads
	High	Highways Piped sewerage and treatment plant	Electric power supply Piped water supply Schools Hospitals
		TOLL OR CLUB GOODS	**PRIVATE GOODS**

Figure 3.3 Degree of subtractability and excludability in selected economic infrastructure.

In contrast, the benefits obtained by any individual consuming a public good is 'externalised' as they become available to all others. Figure 3.3 shows the degree of excludability and subtractability for selected infrastructure facilities.

Club goods are characterised by low subtractability and high excludability. Examples include highways, sometimes referred to as a 'toll system'. Goods such as urban roads that are characterised by high subtractability and low excludability are referred to as 'common property'.

The application of principles of excludability and subtractability has important policy implications in terms of funding at national and local level, and the way charges and fees are paid for by consumers of infrastructure services. Pure private goods can be effectively provided through the market mechanism i.e. fees and charges can be levied directly to the consumers of such services. In contrast to pure private goods, consumers will not voluntarily offer payments to the suppliers of pure public goods. The linkage between the supplier and consumer is broken and therefore needs government intervention in the provision of such infrastructure. This requires a political solution, and not the market mechanism, to decide on the level of infrastructure to be provided, how consumers should pay, how it will be funded in terms of taxation, whether nationally, locally or ring fenced. For example, the benefits of certain infrastructure (local public goods) such as local roads, street lighting, and local parks are spatially restricted and therefore can be subject to local taxation.

The potential for private sector participation, or public sector intervention in infrastructure is not only determined by the degree of subtractability and excludability but by other factors as well. Some of the goods that are traditionally provided by the public sector such as power and water supply have private goods characteristics. This explains why there has been a significant shift in recent years to encourage private sector providers in the delivery of power and water services. Similarly, hospitals, schools, sports and leisure facilities are highly subtractable and highly excludable goods, and therefore have private goods characteristics. The existence and operation of private schools and hospitals all over the world is ample evidence that such goods can and will continue to be provided by the private sector. However, most hospitals and schools are still provided by the public sector as the consumption of such goods by the whole community far exceeds the benefits to individual consumers.

Public service obligations and externalities

Public service obligations and externalities are also crucial in determining the level of public or private participation. Providing affordable minimum access to transport, water and sewerage, education, health, community and recreational services are important public service obligations that cannot be left entirely to market forces. By way of illustration it may be necessary to maintain public transport to certain areas that would be uneconomical or unprofitable for private firms to operate. The reasons for the provision of such a service are many and varied according to local conditions and may include the need to facilitate national integration and to create access to employment opportunities. Externalities (third-party costs and benefits) associated with the development of infrastructure such as air, land and water pollution, adverse effects on public health, noise and safety also justifies the need for public sector involvement, operation and regulation. The implication is that the quality of private infrastructure services must be constantly regulated, monitored for coverage, standards, and safety.

Infrastructure facilities traditionally provided by the public sector (for example, power and water supply, hospitals, schools, sports and leisure facilities) have private goods characteristics. However, Table 3.2 shows a number of public service obligations and externalities associated with the provision of such facilities. Airports are regulated to ensure that noise levels are within acceptable limits that take into account a balance of commerce and community interest. Certain types of land such as rural, recreational and green areas are protected from the development of roads and other infrastructure through the use of planning mechanisms to mitigate against the risk of water, air and noise pollution.

The public sector has an obligation to intervene in the provision of certain types of major infrastructure such as airports, seaports as they are central to the national interest and economic prosperity. Intervention is also important to national security and the fight against national and international crime and more recently terrorism. However, it is vital that adequate personal and institutional support to 'security' infrastructure facilities sometimes referred to as 'global' goods, are provided in the form of immigration control, customs and

43

Table 3.2   Examples of externalities and public service obligations

Infrastructure type	Externalities	Public service obligations
Railways	Network effects	National integration, access to services for remote areas
Roads	Affects settlement/ land use patterns, drainage, erosion, public safety, dust pollution	National integration, access for remote areas
Airports	Noise, public safety	National and international integration, national defence and security
Water supply and sewerage	Public health Water pollution	Affordable access to minimum services
Education	Productivity, public health and safety	Affordable access to minimum services Achieving certain political goals (e.g. national unity, participation in development)
Hospital and health care	Public health and safety	Affordable access to minimum services
Sports and recreational facilities	Public health, noise, public safety	Affordable access to minimum services, local, national and international integration
Factories and warehouses	Air, noise and water pollution Public health and safety	Access to employment
Shopping facilities	Noise, public health and safety	Access to employment
Offices	Noise, public health and safety	Access to employment

policing of national entry and exit points through which people and goods flow. Developed nations have a well established appreciation of the need for 'security' infrastructure, however this has not necessarily been the case in all developing countries. Globalisation has generated the need for international cooperation to improve global security infrastructure in the fight against international crime and terrorism. All countries are now under pressure to increase infrastructure security to comply with internationally accepted practice.

Development and service delivery issues

The characteristics of sub-sectors that exist within major categories of infrastructure are important, since their differences are represented by those infrastructure functions that are potentially competitive and those that are monopolistic. Different policy issues, therefore, arise in competitive and monopolistic infrastructure activities in the development and service delivery phases. Examples are shown in Table 3.3.

Table 3.3 Policy objectives in competitive and monopolistic infrastructure sub-sectors

	Development phase (bid preparation, design and construction)	Operational phase (Service delivery and facilities maintenance)
Potentially competitive sub-sectors	• Designing market structures for public to private transition	• Implementing general competition policy, including network interconnection arrangements for certain subsectors
	• Establishing rules for externalities and other issues • Reducing the public sector's direct role in contracting	• Providing explicit subsidies for basic services for the poor • Providing light touch regulation
Natural monopolies	• Efficiently managing the contracting process using 'competition for the market'	• Regulating the sector to ensure fair pricing, low-cost service delivery, high-quality service and adequate future investment
	• Managing externalities and other resettlement issues	• Providing explicit subsidies for basic services for the poor

Source: Adapted from Kohli, Mody and Walton (1997, p. 8).

In the development phase, the private sector tends to be very concerned about development costs and risks. This concern is influenced by the contracting or bidding process (whether procurement is local or international, competitive or selective), bidding and transaction costs, the granting of development permits or planning permission, externalities such as noise, air, water pollution and other environmental and resettlement issues, and associated public obligations such as employment creation. There are also issues relating to the nature, length and outcome of negotiation, design and construction, contract conditions and service level expectations. Transparency and fair selection processes are essential to the confidence of bidders and this will have a direct impact on risk perception with consequential effect on bids.

To mitigate design and construction risks, particularly for large and complex infrastructure projects, it is necessary for stringent prequalification procedures to select experienced design and construction firms. These risks may be transferred to contractors by entering into appropriate contractual arrangements such as fixed-price and date-certain turnkey construction contracts, with penalties for non-performance. However, bidders will need to be reassured regarding the level of risk and failure; this will result in risks being priced accordingly leading to higher than necessary bids for the project. In extreme cases bidders may withdraw from the bidding process altogether.

Competition and cost recovery

The potential for competition is determined by the level of sunk costs, and the existence of scale or scope economies. Infrastructure activities requiring major capital investment subject to high sunk costs and long payback periods are generally less attractive for the private sector because of the risk involved. The theory of contestable markets shows that the higher the sunk costs, which is a function of both entry and exit costs, the lower the potential for competition. Entry cost depends on the infrastructure facility development (bidding, planning, design and construction cost) and service delivery costs. Exit costs depend on the transaction costs associated with ceasing or closing the operations of a particular service. Asset specificity is a key factor determining the level of exit costs. To illustrate this point the exit cost of a private railway company will be high since it may be able to sell rolling stock to other companies, but the rail-track and its associated infrastructure must stay in place and will be valued accordingly. This is especially the case when some parts or the entire network is unprofitable. The British Government dealt with this problem by separating the ownership of the rolling stock from that of the permanent track infrastructure and its associated management and maintenance (Crompton and Robert, 2003).

The existence of economies of scale or scope means that competition can be created by vertical or horizontal unbundling of the activities in any particular infrastructure sector. For example, power infrastructure can be divided into generation, transmission and distribution by vertically unbundling the network, whereas railways may be divided geographically by horizontal unbundling with different operators serving particular regions, and where appropriate the separation of the railway infrastructure (tracks) from the rolling stock. The potential for cost recovery will also affect the level of participation by the private sector. Large urban piped water supply networks and power transmission grids are intrinsically monopolistic and they provide private goods with a high potential for cost recovery from user charges. Clearly, this situation will attract public intervention to maintain the best interest of users resulting in regulated services, prices and acceptable levels of public safety. Other factors include willingness to pay, income levels and the availability of substitutes.

Case studies

Case study 3.1: The UK Private Finance Initiative (PFI)

In 1992, the UK government faced with significant capital constraint in the public sector and the need to remain competitive in the global market introduced a hybrid approach called the PFI combining public and private funding to deliver infrastructure projects. This approach reflected a radical shift transferring the responsibility for design, construction and operation of infrastructure facilities to the private sector (Grout, 1997).

The policy objective of PFI is to improve the level of public services and was underpinned by a theory focusing on the delivery of services rather than the construction or ownership of infrastructure assets. A key challenge for the private sector was understanding their service obligations with respect to the performance of the infrastructure assets for say 20–30 years. The implications are profound for bidders in terms of managing the balance between the capital costs and operating costs to achieve acceptable service performance levels unlike traditional procurement where responsibility for infrastructure assets ceases after the defects liability period usually a year after construction is completed. This requires knowledge and a significant input of facilities management or operational aspects at the early stages of design.

A legislative framework was established to promote the use of the PFI model. Both institutions and actors involved were identified. This included local authorities, National Health Service Trusts, government departments and agencies, financiers, private construction organisations, financial, legal and technical advisers. The resource implications were also identified such as government funding, the level of private sector interest, their capacity and willingness to enter into partnership with the public sector. The information machinery necessary to explain roles, contractual relationships and responsibilities in the new procurement approach was crucial. A number of documents and publications were produced and continuously reviewed by the government and support agencies on how to put PFI theory into practice. For each stage, specific guidance notes were made available to review the potential for PFI within public sector institutions and various other sectors. Market testing was implemented to ensure private sector interest and the viability of projects. The business need for a service was tested and project options were explored to demonstrate value for money. Information and feedback from earlier projects were also made available to develop best practices.

There was a delay in the initial take up of PFI schemes by the public sector due to a number of reasons ranging from legal, sustainability to competition issues. In health specifically there was a problem with the risk of local substitution after major PFI investment because of the internal market created. To date over 20 government departments have been involved in PFI schemes with significant activities in the health, education and transport sectors, prisons, fire stations, housing and waste management.

Author's commentary
The PFI reflects a radical change of policy in the UK aimed at encouraging private sector finance and management expertise in the delivery of traditional public services in health, education, transport and other key areas to make the UK economy more competitive. PFI is a 'buy now-pay later concept' and like all credit agreements there is interest to be paid. However, the policy outcomes are significant and the benefits in terms of whole life performance of infrastructure facilities and improvement in the delivery of public services are felt in hospitals, schools and many sectors across the UK. A total of 451 PFI projects were successfully completed by 2003 providing over 600 operational facilities amounting to £666.8 million. By April 2003 a total of 63 PFI had been transacted, resulting in a total capital value of £35.5 billion (HM Treasury, 2003).

Case study 3.2: Australian public funding of infrastructure: changing the model

Establishing needs and setting priorities has in the past, facilitated the provision of public infrastructure in Australia and projects have been primarily procured traditionally by in house management, a public sector workforce and capital provided by taxation or bond issues. Today there are a number of options available, ranging from the traditional approach to complete privatisation and outsourcing. There has also been devolution of responsibility from federal government to state governments, typified by 'untied' budget allocation and a shift in responsibility to provide facilities and services locally.

A trend is emerging that provides states with powers and responsibility to supply important infrastructure that in some cases might have significant national importance. This raises the issue of a national long-term infrastructure strategy necessary to provide direction as well as facilitating agreement on an integrated approach to the provision of infrastructure between state governments. In this manner inconsistency, duplication and fragmentation can be eliminated or reduced.

A national transport plan called 'Auslink' has been instigated by federal government that recognises the need for strong national focus, while accepting that states will have much more direct responsibility for infrastructure provision. Auslink provides for a National Transport Advisory Council consisting of experts from the private and public sectors who provide transport ministers with strategic analysis and advice on Australia's transport provision. In addition to providing strategic direction Auslink encourages effective partnership between national and local governments and communities in the delivery of efficient and needed transport using new technology where appropriate. By this means a more strategic local government approach is supported by broad funding that takes on board national infrastructure strategy.

Within the context of Auslink there have been some promising developments at the state level. The State Government of Victoria has appointed an Infrastructure Planning Council to examine the state's infrastructure requirements over the next 20 years in the areas of energy, water, transport and communications. Similarly, the New South Wales State Government released in 2002 a 10-year plan for renewing the state's capital base. In Queensland State an infrastructure plan was produced in 2001 that supports a comprehensive and integrated approach to long-term strategic growth and investment, as well as providing a better service for users.

Author's commentary
A major concern is that states within a federated system of government will use devolved responsibilities to challenge or not accept national plans and strategies. The key issue revolves around the degree of autonomy that is handed down and the extent to which central government applies sanctions and limitations in funding resulting from non-compliance with federal strategies that are in the national interest. A productive way forward might be the development of partnership and teamwork between federal and state governments aimed at facilitating the development of a truly national and co-ordinated infrastructure strategy that provides a framework within which local infrastructure can be developed.

Summary

The chapter has illustrated the critical need for establishing and articulating a policy framework to improve the efficiency and effectiveness of infrastructure delivery strategy and the sustainability of infrastructure assets. It is argued that the environment, institutions, goals and objectives, resources, knowledge, information and communication systems are central to the policy-making process, and the formulation and implementation of appropriate infrastructure policies requires an understanding of these key elements and their interactions. The key economic factors that determine the level of public or private participation in the delivery of infrastructure are discussed. However, it is noted that decisions on whether the private or public sector should provide a particular type of infrastructure are rarely based on economic considerations alone but a range of other non-economic factors such as public service obligations, externalities, national security and defence central to making policy decisions. The chapter concludes with two case studies illustrating the principles discussed. The first case, the UK PFI, demonstrates that radical policy changes can be successfully implemented, if supported by clear goals and objectives, the identification of institutions, resources and an information and communication systems to support implementation.

The second case focussing on the Australian situation recognises the need for strong national focus in policy making and implementation, while accepting that local governments or states will have much more direct responsibility for infrastructure provision. It is argued that effective partnership between national and local governments and communities is the key to the effective delivery of infrastructure projects.

References

Broadbent, J. and Laughlin, R. (2003), '*Control and legitimation in government accountability processes: The private finance initiative in the UK*', Critical Perspectives on Accounting, 14, 23–48.

Crompton, G. and Robert, J. (2003), '*Such a silly scheme: the privation of Britain's Railways 1992–2002*', Critical Perspective on Accounting, 14, 617–654.

Grout, P.A. (1997), '*The economics of the private finance initiative*', Oxford Review of Economic Policy, 13(4), 53–66.

HM Treasury (2003), '*PFI: meeting the investment challenge*', online at www.hm-treasury.gov.uk

Jenkins, B. (1993), '*Policy analysis: models and approaches*'. In A. Reader, M. Hill (eds), The Policy Process; Harvester Wheatsheaf, Hertfordshire, UK, pp. 34–44.

Kessides, C. (1993), '*Institutional options for the provision of infrastructure*', World Bank Discussion Papers Number, 212, World Bank, Washington, D.C.

Kessides, I.N. (2004), '*Reforming infrastructure: privatisation, regulation and competition*', A World Bank Policy Research Report, World Bank, Washington, D.C.

Kohli, H., Mody, A. and Walton, M. (1997), '*Making the next big leap: systemic reform for private infrastructure in East Asia*', In H. Kohli,

A. Mody and M. Walton (eds), Choices for Efficient Private Provision of Infrastructure in East Asia, World Bank, Washington, 1–20.

Miller, J.B. (2000), '*Principles and practice of public and private infrastructure delivery*', Academic Publications 2000, Kluwer.

Morah, E.U. (1996), '*Obstacles to optimal policy implementation in developing countries*', Third World Planning Review, 18(1), 79–105.

Ostrom, E., Schroeder, L. and Wynne, S. (1993), '*Institutional incentives and sustainable development: infrastructure policies in perspectives*', Westview Press, Boulder.

World Bank (1994), '*World development report 1994: infrastructure for development*', Oxford University Press, New York, USA.

Chapter 4

Strategic considerations for infrastructure development

Introduction

Infrastructure programmes are needed to achieve a range of development objectives focusing on issues such as raising standards of living and welfare, improving access to transport, education and health services, facilitating economic development and creating employment, as well as promoting environmental sustainability. There is a need for strategies to deliver these development objectives. Network infrastructure such as roads and highways, railways, electric power, water supply and sewerage systems needs long-term strategic planning due to the inherent difficulties in co-ordinating large and complex networks. There is also a strong argument for public sector intervention in the planning of airports, ports, waterways due to national security, and in some other cases there are public service obligations for the provision of services such as schools, hospitals and solid waste management. These services are vital in protecting the citizens of a country and in providing a galvanising effect to facilitate social cohesion and economic development.

There are also problems of externalities (e.g. public safety, pollution, public health, erosion etc.) often associated with certain types of infrastructure development or lack of it. Furthermore, the market failure argument requiring the control of land resources for infrastructure development means that public sector intervention will always be required in the formulation of local, structure and master plans to guide development activities. These factors mean that the public sector role is critical in infrastructure delivery to facilitate co-ordination, to achieve development objectives such as national security and public service obligations, and to minimise the effects of market failure and externalities. Infrastructure provision cannot therefore be left to the invisible hands of market forces. A co-ordinated planning approach is widely supported by policy makers and planners but infrastructure development processes pose major challenges for the actors and institutions involved in its delivery. The key issue for national governments is how to transform policies into effective strategies for the implementation of infrastructure projects and the maintenance of infrastructure assets to improve the delivery of public services.

The previous chapters have discussed the characteristics of different types of infrastructure, and the need for developing a coherent policy framework to facilitate infrastructure implementation. This chapter explores the strategic implications of a number of themes running through the earlier chapters. Following the introduction, the key elements influencing the formulation of a strategy, together with the level of infrastructure development are outlined. Also discussed is the complexity of the infrastructure development process and the need for a holistic approach to identify systemic failures in infrastructure using a variety of systems techniques (Checkland, 1991; Flood, 1995; Yeo, 1995). The infrastructure delivery chain underpinned by the concept of value added to identify bottlenecks and to continuously improve infrastructure services is presented. It is this service-focused approach that is central to the sustainability of infrastructure and the delivery of services. Finally, the need for a decision-making hierarchy to facilitate the management of resource demand and supply to improve infrastructure services is explored.

Key elements of infrastructure strategy

There are three major thrusts to the development of a strategy for infrastructure delivery. First, there is a need for infrastructure investment to address existing constraints and to provide a conductive environment to facilitate and co-ordinate investment requirements. Second, knowledge resources or a range of expertise are required to identify gaps in the provision of infrastructure and then to plan, structure and develop appropriate projects. Means must also be established to enable the implementation of projects and maintain them efficiently. Third, it is vital to have a planning and regulatory framework to deliver development objectives in a sustainable way, and to provide resources to review infrastructure projects, monitor their implementation and to enforce appropriate service delivery standards.

Investment issues

The demand for infrastructure is derived from household, commercial and industrial needs and expectations. This has been increasing in almost all countries and is expected to continue (Fay and Yepes, 2003), although the composition of demand varies between different countries at various levels of development. The levels at which these demands are fulfilled is a function of the strategic framework to co-ordinate investment requirements. To meet the rising demand for infrastructure, issues of investment should be addressed that stem mainly from the public sector or the involvement of the private sectors in areas that are likely to be profitable.

Public sector budgetary constraints affect the level of development investment and recurrent expenditure. Both the theory and practice of development policy suggest that infrastructure investments are a major factor in economic growth, but productivity of infrastructure assets has been undermined where there are constraints in public sector recurrent expenditure (Kalaitzidakis and Kalyvitis, 2004). The implication is that the lifecycle of infrastructure facilities are significantly reduced with enormous costs and consequences to governments and the private sector. Low infrastructure service pricing due to non-economic considerations, such as equity and public service obligations sometimes underpinning pricing policies also causes constraints in public sector financing. User charges do not always reflect the true cost of infrastructure provision but the principles of cost recovery are fundamental in guiding infrastructure financing decisions, to ensure adequate maintenance and the sustainability of services. The potential for private investment will therefore increase not only as public sector expenditure and borrowing continues to be tightened, but as infrastructure schemes become more commercially viable.

As a result of limited public sector budgets, and in some cases severe constraints, private sector contribution in infrastructure is gradually becoming important (Esfahani and Ramirez, 2003). The question of how to increase infrastructure investment from the private sector has now become a major challenge (Kessides, 2004). There are two main sources of private sector finance – domestic finance through national development banks, and other private sources or external finance from international commercial banks and other private sources from abroad. It is increasingly recognised that there is a need for private finance in infrastructure. Recent trends towards public–private partnerships reflect the view that the private sector can make a substantial contribution towards development (Miller, 2000). For example, the UK Government introduced the Private Finance Initiative (PFI) in 1992 to facilitate private investment and management of infrastructure facilities to improve the level of public services. Through PFI, the responsibility and risk of developing, operating and maintaining key facilities in healthcare, education, transport, defence sectors has been transferred to the private sector for typically periods of 20–30 years. There is also recognition of increased private sector role in increasing infrastructure investment in many other developed and developing countries. According to the International Finance Corporation (IFC), developing countries will require more than $3 trillion for investment in new infrastructure over the next 10 years. However, the scope for private investment in infrastructure depends on implementing economic and political reforms, supported by appropriate strategies, to improve investors' interest, commercial returns, risk management and uncertainty.

Knowledge base and resource considerations

Investment constraint is not the only problem affecting infrastructure development. Knowledge required to assess infrastructure gaps or needs, and project structuring and development expertise are crucial; so are the design, construction and facilities management skills necessary for project implementation and maintenance of infrastructure facilities. Expert knowledge is required to conduct sectoral assessments to provide the basis for identifying the potential and constraints in infrastructure development. There are several types of assessment techniques. Technical assessment involves detailed engineering studies of the condition and performance of existing infrastructure stock and appraises the need for replacement. This is usually accompanied by economic appraisals based on the expected costs and benefits of the projects to determine rates of returns. Other methods include econometric and political assessments. Econometric assessments are based on macro data that establishes relationships between rates of growth of infrastructure investment, national income and industrial productivity. Political assessments are often not based on infrastructure gaps but on a desire to influence likely political or voting outcomes, as certain infrastructure projects tend to have a 'high visibility impact'.

Limited project development expertise is also a serious constraining factor that influences the level of infrastructure provision. A well-structured project striking an appropriate balance between debt and equity finance, public and private finance is essential in attracting investors. The proportion of equity to debt financing depends on the infrastructure sector. Ahluwalia (1997; p. 4) noted that telecommunication projects with relatively high market risks require a relatively low-debt component (debt to equity ratios close to 1:1), while power projects with assured power purchase agreements would be financeable with debt to equity ratios of 2.5:1 or even 3:1. This process of project structuring is becoming increasingly complex, particularly for infrastructure projects with long payback periods and high risks that often require financing from several sources. Capacity for project structuring and development is crucial as projects to be funded by investors are often scrutinised for economic returns, risks, project sustainability and consistency with sectoral policy priorities and country strategy. However, project structuring expertise is not readily available in the public sectors, particularly in developing countries. The implementation of the World Bank AGETIP model (Pean and Watson, 1993) in a number of African countries was aimed, partly, at encouraging project development capacity and procurement at central, local government and community levels.

Developing an appropriate project structure and securing investment are not sufficient to improve the delivery of infrastructure services. Economists argue that even if a country has succeeded in raising investments, it takes decades to transform master plans and projects into infrastructure capital – roads and highways, power plants, rail networks, factories, water supply systems – that underpins a productive economic structure. The amount of infrastructure capital or infrastructure facilities constructed depends to a large extent on the capacity of construction sector. Construction is a complex process involving the organisation of diverse resources, and it is the composition and mobilisation of these resources, which determines the level of development and rate of implementation of infrastructure projects (Raftery et al., 1998). There is a range of construction resources required for infrastructure development – plant and equipment, skilled workers and construction materials. Lack of skilled personnel and various resources affects the implementation and productivity of infrastructure. Sir John Fairclough in his report on 'Rethinking Construction Innovation and Research' for UK construction industry noted that there is considerable concern in the supply of professional skills required in the built environment as a result of the dramatic decline in new entrants in construction-related degree courses (Fairclough, 2002).

During periods of high economic activity the capacity problem becomes acute as it takes years to develop certain resources. Typically up to 7 years may be required to produce fully qualified architects, engineers, surveyors, planners and construction and facilities managers. The training of craftsmen or apprenticeships for plumbers, masons, electricians, bricklayers, plant operators, and carpenters takes from 2 to 5 years. Significant timescales may also be required to develop or address problems in manufacturing capacity and to improve the supply of building products, materials, construction plant and equipment without adversely affecting the economy. The cyclical nature of infrastructure investment means that the adoption of casual employment practices by construction organisations is widespread. The implication is that there are often large number of small firms unwilling or unable to grow, and reluctant for their staff to undergo long-term training in construction. Only the larger contractors tend to be able to attract and maintain a high-calibre workforce, as they are most likely to offer structured employment and training programmes.

Planning and regulatory issues

Planning and regulatory functions are intrinsically linked. At the front end there is a need for planning to ensure the identification and implementation of appropriate infrastructure projects, while at the back end, there is the regulation of services derived from the operation of infrastructure facilities. Planning is essentially a political process, notoriously complex at times as a result of the diverse interests of different stakeholders. Conflicts often arise between political and professional planners, local communities, developers and special interest groups. While developed countries tend to have a transparent, more structured planning processes and enforcement regimes, there are serious constraints in developing countries (Brown and Wolfe, 1997; Wells, 2001). For example, the UK planning system is often criticised as archaic, and is considered to be responsible for serious delays in development and construction of infrastructure projects. Such planning practices resulting in lengthy procedures and consultation with stakeholders, often with conflicting interests, are usually interpreted as anti-development or what is sometimes referred to as 'not-in-my-backyard' syndrome. This is in sharp contrast to the opaque planning systems in many developing countries dominated by politicians at national and local level because of the enormous implications and visibility of infrastructure development. Greater reliance on professional planners is necessary to make planning and development control more efficient.

Development plans provide the basis for implementing infrastructure projects at national and local government levels. This includes both economic plans and physical or land use plans, which constitute the means by which governments specify the type of infrastructure, where it is required, and how it allocates and manages its resources. Physical planning concerns the identification and allocation of land for specific uses such as the location of social, trade and technical (economic) infrastructure. Planning laws deal with the regulation of development activities, co-ordinating local, structure, regional and master plans, evaluating planning applications and monitoring development activities. It includes guidelines for infrastructure standards such as heights, lengths, widths, areas, volumes, structural and aesthetic aspects, performance, restriction of specific materials, plant and working methods and hours. Planning processes are aimed at balancing the demands for infrastructure projects against the conflicting interests of different stakeholders at national and local levels. It is these planning processes that are crucial for the implementation of infrastructure projects, to ensure adequate quality, reliable services from infrastructure facilities and to achieve other development objectives. For example, in the UK section 106 planning agreements are used to achieve certain socio-economic development objectives such as urban regeneration, employment creation and skills development programmes for local disadvantaged people and communities, and the integration of local small medium enterprises (SMEs) in the construction supply chain (Howes and Robinson, 2001).

A regulatory framework for monitoring the standards of infrastructure facilities and service delivery is also essential. Regulatory standards tend to focus on balancing quality and pricing issues necessary to protect the environment, consumers and other stakeholders. This often involves licenses, investment issues, service agreements based on pricing and performance, bonuses and penalties associated with non-compliance with health and safety, public health, pollution, noise and other environmental standards leading to market failure. Regulatory mechanisms can counter market failures. However, each infrastructure sector has different considerations reflecting development priorities, efficiency and equity objectives. There are also various institutional options for carrying out regulatory functions. Some regulators have remained or tend to be within government or related agencies, although the situation is gradually changing. Consumers of infrastructure services have become increasingly reliant on independent, either single-sector-specific, clustered or multisectoral regulatory agencies in water, utilities, telecommunications, power and transport sectors. Examples of sector-specific regulatory agencies can be found in the UK (Ofwat, Network Rail). Clustered regulatory agencies are responsible for a few

57

related infrastructure sectors such as an energy regulator for gas and electricity and a communication regulator for posts and broadcasting. Multisectoral regulatory agencies responsible for most infrastructure sectors include states in the USA. These are particularly useful in developing countries with limited resources, regulatory capacity and experience.

Infrastructure development processes

Increased globalisation means that the level and quality of infrastructure development has become more critical than ever for national economies to remain competitive and innovative. Infrastructure development is a complex process influenced by a variety of factors. It is the contextual variables of a country such as economic, social, environmental and political factors, which interact to determine the composition, priorities and timing of infrastructure programmes. Examples of selected factors and their relationships with certain variables are shown in Figure 4.1.

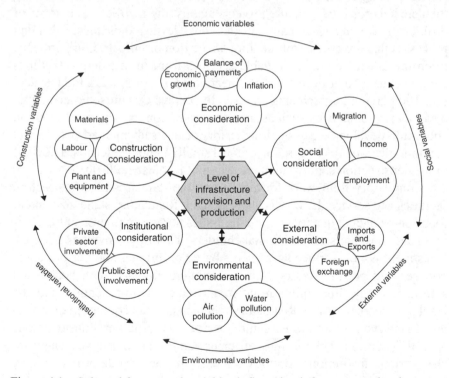

Figure 4.1   Selected factors and variables influencing infrastructure development.

The variables impact on these factors to influence infrastructure delivery such as the availability of materials and skilled labour, balance of payments, employment, imports and exports, pollution, and private sector involvement, and land use policies. Some of these variables are internally driven while others are externally imposed. However, it is the contextual

58

variables that help to shape the investment environment, knowledge base and resource capacity, and the planning and regulatory framework so vital in the development of a strategy for infrastructure development. These contextual variables interact in a complex and dynamic way and are not easily understood using traditional fragmented or 'hard' approaches. A holistic approach is necessary to ensure that the complexity of the infrastructure delivery process is understood, and more importantly, to help identify specific institutional factors, actors and resources that could facilitate infrastructure development.

The need for systemic planning philosophy

An understanding of the infrastructure delivery process is crucial in order to be able to identify the determinants of infrastructure demand and supply, and their interactions. For example, the construction of offices, factories, shops, business parks will lead to an increase in telecommunication and energy demand. The production of local building materials to fulfil infrastructure demand depends on natural resource endowment. However, building materials production is an important contributor to environmental degradation. The rapid growth of sand mining and quarrying activities to meet the demand for construction projects in some countries is a source of major environmental problems. Increased road construction and use also affects the demand for railways and construction resources. Investment in hotels and related tourism infrastructure to attract foreign tourists could be accompanied by an airport expansion programme. A massive investment in export free zone facilities for factories and business parks may be accompanied by an expansion of airport and seaport facilities. Such linkages are crucial in understanding the demand and supply implications of infrastructure projects. A classic example is the case of Nigeria during the oil led construction boom. Cement was left offshore in ships and eventually deteriorated because of limited capacity in Nigerian ports. The implication was that the same cement was often bought and sold offshore with large profits for some and larger losses for others and as a result many projects were not completed (Hillebrandt and Meikle, 1985). It is these systemic effects that are too often ignored, underplayed or at best inappropriately captured in the understanding of construction processes (Howes, 1996), and the development of an infrastructure strategy due to the dysfunction of government agencies responsible for co-ordinating infrastructure and associated investments (Robinson, 2000).

Problem structuring methods

The elements in infrastructure development need to be structured to facilitate systemic planning. Key tools used in systemic planning are system techniques for capturing the essential aspects of a problem situation and structuring it for investigation (Checkland, 1991). Systems diagrams are powerful devices for synthesising and analysing problems. Examples include decision diagrams, survey graphs, systems maps, influence diagrams, cause-and-effect diagrams and cognitive maps. Although there are a variety of system diagrams, some methods are more suitable than others for tackling particular types of problems. For example, influence diagrams are more useful than decision diagrams in situations where events are not independent of each other. Also, influence diagrams perform better than decision diagrams on unstructured problems, and are therefore used for the 'most critical of decisions such as strategic planning decisions'. A collection of these techniques can provide decision-makers with powerful tools for structuring problems of infrastructure development, and to capture a range of issues and context. The following section provides examples of the application of some of the techniques in the context of infrastructure development.

'Rich picture' of infrastructure system

A country's infrastructure system does not exist in isolation. Programmes for infrastructure development often reflect socio-economic, environmental and institutional objectives. Educational infrastructure is required to improve access, boost enrollment, and in the end to create a more educated and productive workforce.

Similarly, social and health infrastructure contribute directly to the improvement of living standards, and indirectly to industrial productivity. Trade infrastructure is required to support business services and production. Power supply facilities are needed to support households and businesses. Better roads, seaports and airports are vital to improve transportation, international trade and industrial activities. Figure 4.2 is an example of a 'rich picture' of an infrastructure system.

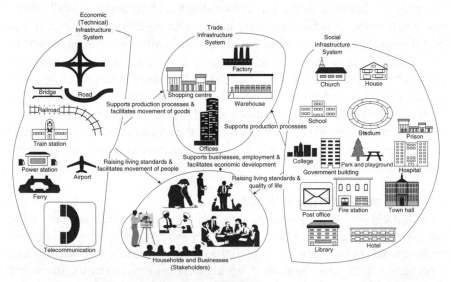

Figure 4.2   'Rich picture' of the infrastructure system.

Survey graph

The 'rich picture' of the infrastructure system shown depicts different types of infrastructure, their functions and relationship with other facilities and stakeholders. However, the 'rich picture' is not comprehensive as it does not show how infrastructure facilities are actually built. The processes, resources and actors involved in infrastructure delivery are completely ignored. To improve the rich picture further, various resources required could be included. Figure 4.3 is an example of a survey graph detailing some resources required for infrastructure production.

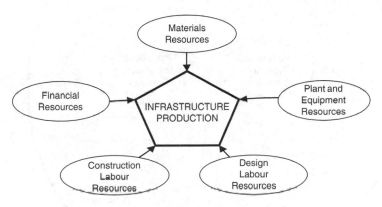

Figure 4.3   Survey graph for infrastructure production.

Unfortunately, both the rich picture and the survey graph are insufficient for understanding the dynamics of infrastructure development. They do not specify the activities involved nor do they include resource attributes such as quantity, quality, cost, etc. Also, infrastructure activities are performed by different actors whose behaviours are influenced by socio-economic, institutional, environmental and other factors.

Systems map

Additional information may be added to the survey graphs and synthesised to form a 'systems map'. Figure 4.4 is an example of a base-level systems map. It shows not only the construction inputs (resources) necessary for infrastructure production, but also the actors and other resources such as 'planning and regulatory' resources indirectly supporting infrastructure development.

The systems map specifies the different components of the infrastructure system. The missing components not shown on the survey graph such as the actors – planning, infrastructure organisations, manufacturers, design/engineering organisations, construction organisations – and other issues relating to infrastructure standards, development control standards, and design and technology are clearly identified. The systems map shows the relationship between key factors and the strategic variables of the system.

In practice, it will be necessary to break the systems map down further into more detailed subsystems within those already shown.

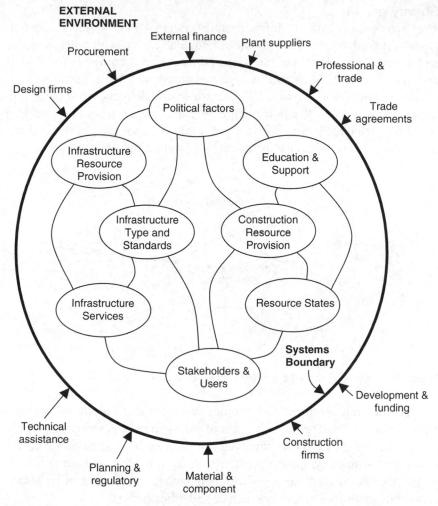

Figure 4.4   A systems map for infrastructure resource development.

Infrastructure service delivery chain

To allow a further examination, an infrastructure systems map is transformed into an infrastructure service delivery chain to explore important relationships between the components and their elements. Failure to focus on the service delivery chain and to take account of sequencing and weakest links in the chain is one of the main reasons why infrastructure projects have not had the desired or intended effects. The service delivery chain enables a refocusing of infrastructure development beyond the realisation of the construction phase, by extending it to the use of the end product and its maintenance. Through the use of the 'delivery chain' (see Figure 4.5), systemic failures can be identified and value added by incorporating or improving the state of relevant institutional actors and resources at key process stages to improve the delivery of infrastructure services.

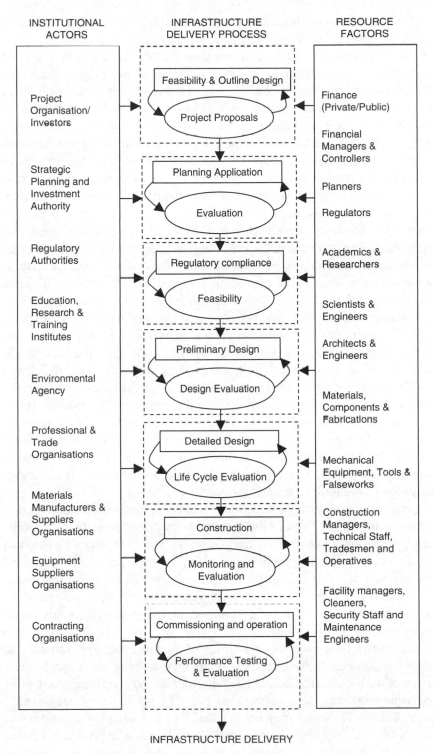

Figure 4.5 Infrastructure delivery chain.

The value chain is a logical expression of infrastructure delivery, consisting of forward and backward interactions, influenced by the strategic variables of institutional actors and resource factors. Infrastructure project proposals usually mean increases in the number of planning applications. The greater the number of planning applications submitted for evaluation, the more the need for planning resources to assess regulatory compliance with local, structure and master plans as well as other development objectives. An increase in the number of planning applications approved could mean increases in design and construction activities, which in turn creates an increase in planning resources to ensure that proposed development activities comply with development control standards. Increased construction activity also creates an increase in the demand for design and construction personnel (architects, engineers, construction managers, etc.), materials, plant and equipment. The state variables reflecting the stages in the infrastructure development process could be stable or chaotic depending on whether appropriate resources and institutional actors required are available. An unstable situation could arise if there were too many infrastructure projects, but limited planning, design and construction capacity to cope with implementation.

The concept of value is fundamental to achieving continuous improvement in infrastructure delivery. There are three conditions required to add value at each stage of the infrastructure delivery process in order to achieve a satisfactory outcome. These are the relations of (a) necessity, (b) sufficiency and (c) requirements. To achieve the development state of 'infrastructure project is viable', the following conditions of necessity are required: (1) a *project owner/organisation*, (2) *a project proposal*, (3) *finance* and (4) *planning permission*. But none of these conditions alone would be sufficient. The condition of sufficiency captures all necessary conditions to achieve a satisfactory outcome.

Similarly, to achieve the development state 'infrastructure construction is satisfactory' (see Figure 4.6), the necessary conditions are – (1) *infrastructure design development is satisfactory*, (2) *employ construction labour*, (3) *obtain materials*, (4) *obtain plant and equipment* and (5) *infrastructure production standards are satisfactory* i.e. *project has complied with standards and regulatory requirements*. However, a third condition – the *relations of requirements* – is vital to reflect the qualitative aspect of the institutional actors and resources available.

One could have sufficient quantities of all the necessary resources to achieve a particular infrastructure development state but the resources may not be of the required standard. Materials available may be substandard, plant and equipment may be available, but not efficient or in proper working order. Labour resources may be abundant as in developing countries but skill levels are very low. The N, S and R conditions address issues beyond quantitative measures to deal with qualitative factors such as the disposition of institutional actors and quality of resources. As Morah (1996) noted 'an actor's negative disposition can cause them to defy programme objectives by

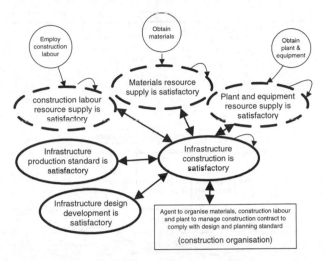

Figure 4.6   N, S and R conditions for infrastructure construction development.

surreptitious diversion and evasion (through oversight or slight enforcement)'. If planning authorities do not adhere to compliance standards the quality of infrastructure service delivery will be compromised. It is therefore important that all *necessary, sufficient and required* (N, S and R) conditions are fulfilled to add value to the infrastructure delivery chain. The absence of any condition will undermine the functionality of the chain and efficient service delivery.

To extend this value-added concept further, different possibilities reflecting various institutional actors and resource options for improving the outcome could be included. For example, financial resources could be obtained from the private sector where there are budgetary constraints in the public sector. Design and construction organisations could be from abroad where there is limited local capacity. Infrastructure projects may be owned by the private sector and facilities/services operated or managed by a foreign company where there are constraints associated with domestic firms. Regulatory compliance activities could also be the responsibility of private sector organisations as is the case in some countries where there are serious problems of capacity in the public sector.

Infrastructure decision-making hierarchy

Designing a hierarchy is a suitable way of structuring a complex problem. Three levels of decision-making can be distinguished with the strategic components at the core of the hierarchy as shown in Figure 4.7.

This hierarchy is not necessarily an exhaustive one but could reflect different worldviews. A view of the operational level shown reflects the view for infrastructure organisations. Another view of the operational level could be shown for planning, materials manufacturers and suppliers, construction and design organisations.

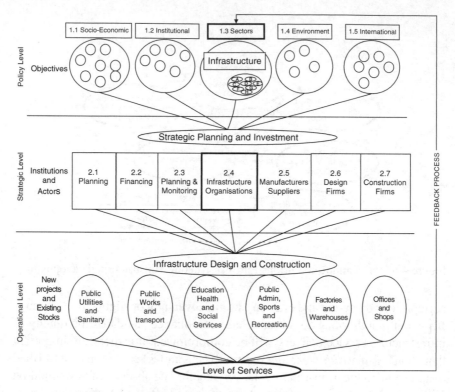

Figure 4.7   Decision-making hierarchy.

There are interactions within levels (horizontal interactions) and between levels (vertical interactions). Context variables such as socio-economic, sectoral, institutional and environmental considerations influence the development of infrastructure strategy. The strategic level deals with infrastructure investment, planning and construction resource provision. The emphasis at the operational level is the transformation of infrastructure projects into facilities and existing stocks for use in the delivery of infrastructure services. The levels of infrastructure and resource provision at the strategic level are influenced by policy factors. For example, private financing for infrastructure is determined by a country's economic and institutional policies. Economic policies supporting privatisation in infrastructure have been shown to improve the credibility of countries on the international financial markets. Similarly, the level of infrastructure service at the operational level depends on the strategy for transforming infrastructure plans and projects into facilities, and their maintenance. The delivery or improvement of particular types of infrastructure services could be hampered by a number of operational problems such as design and construction options, cost overruns, incomplete infrastructure projects due to lack of resources or lack of maintenance of existing stocks. The operational level forms the basis for evaluating the impact of strategies, which in turn informs policy development through the feedback process.

Level of service indicators

Several types of input and output indicators could be used such as performance indicators reflecting service quality, coverage, disruption, satisfaction and willingness to pay. A special facility questionnaire could be used to capture detailed information on the location and quality of infrastructure facilities such as schools, healthcare, roads, water, power supply, recreational and other types of infrastructure. An example of a set of infrastructure input indicators is shown in Figure 4.8.

Infrastructure subsector	Example of selected indicators
Water supply facilities	proportion of dwellings with piped water supply volume of water supplied per capital distance / travel time to wells
Health facilities	number of health centres per thousand inhabitants distance / travel time to health centres
Housing facilities	number of households per dwelling ratio of formal and informal housing proportion of permanent housing structures
Recreational facilities (e.g. parks and playgrounds)	recreational area per inhabitant distance / travel time to recreation facilities
Roads facilities	primary roads (km) per square kilometer secondary roads (km) per square kilometer travel time
Educational facilities	number of classrooms per school population average distance to educational facilities
Power supply facilities	proportion of dwellings with electricity per capita

Figure 4.8   Examples of infrastructure input indicators.

The underlying assumption is that there exists a strong positive correlation between input indicators reflecting infrastructure facilities and output indicators measuring the actual level of services.

Examples of planning indicators include the time taken from submission of planning applications to approvals, time taken from approvals to start on-site. Construction indicators include shortages of materials, delay in design and construction activities, labour and plant shortages. These infrastructure or infrastructure-related performance indicators serve as feedback mechanisms and help to establish priorities on which indicators need to be improved. Problems experienced at the operational level forms the basis for developing strategies to improve resource demand and supply management.

Demand and supply management issues

The decision-making hierarchy provides an insight into the effect of various *policy* objectives on improving *strategic* outcomes for solving *operational* problems. Addressing complex resource problems of infrastructure delivery require a holistic approach incorporating a variety of 'hard' and 'soft' methods (Figure 4.9). The predominant methods for tackling mechanistic problems at operational level are 'hard'. Hard methodologies of operational research such as network analysis, cost optimisation and stock control are used for optimising construction project resources. At this level there is usually a complete consensus on what needs to be done and what constitutes efficiency in doing it. What is needed in terms of construction manpower or energy supply may be easily quantified. However, at the strategic level, how to increase construction manpower or achieve the amount of energy required is more difficult as various resource development or energy options need to be explored to achieve manpower or energy targets.

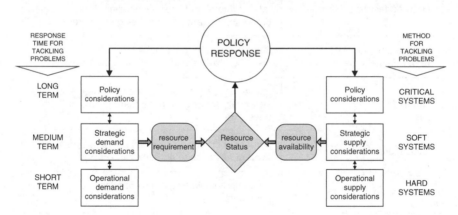

Figure 4.9   A holistic approach to resource management.

At the strategic level, problems become less well-defined (softer) and consensus quickly breaks down reflecting different methods, views and arguments. There will be issues created by clashing norms, cultural values, power domination and one's viewpoint (Figure 4.9). A combination of both hard and soft methodologies such as qualitative analysis are required for higher level unstructured problems, and the task required at this level is to conceptualise the systemic effects. A soft problem can sometimes be transformed into hard problems through structuring and vice versa.

The overall system performance will depend on the type of problem (whether hard or soft), the methodology used in tackling the problems and how well decisions at all levels are integrated.

Strategic hard and soft variables

Infrastructure projects influence the level of resource demand. Resource requirements are affected by more defined hard variables such as design, technology, infrastructure type, investment levels, location and implementation period, see Figure 4.10 showing the relationship between strategic resource (hard) and institutional (soft) variables. The structure of demand has an influence on the structure of supply such as the organisation of the

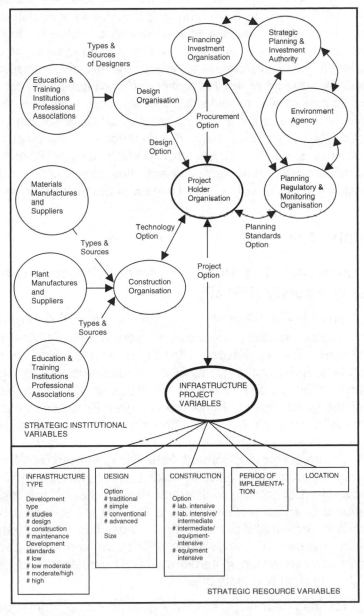

Figure 4.10   Relationship between strategic institutional and resource variables.

69

construction sector. The structure of supply is determined by the capacity of the production units to provide services in relation to the type, size and locational distribution of demand. Nevertheless, improving the supply situation through the availability of resources, is influenced by soft variables.

Resource profiles during infrastructure development are influenced predominantly by soft variables. The procurement route is not simply determined by the 'project holder organisation', but is often influenced by 'investment organisations' or funders. Design and technology choice could be indirectly influenced by 'planning and regulatory organisations'. Similarly, the project type is not solely determined by 'project holder organisations' but is strongly influenced by 'investment organisations' providing funding. Soft variables can alter the way in which infrastructure projects are procured, funded, designed and constructed in accordance with planning and regulatory standards. The structure of demand could be changed through procurement practices limiting the size of tender packages to certain levels as in the UK PFI procurement. The treatment of these variables as 'soft' is not only consistent with the lack of empirical knowledge about the exact quantitative relationships, but is recognition that their influences on hard variables are crucial. In fact, a model that recognises these variables, although soft, is far better than those models that include only a limited number of quantitative variables.

Case studies

Case study 4.1: The Public–Private Infrastructure Advisory Facility (PPIAF)

Recognition of the limitation of existing ad hoc infrastructure development strategies prompted the establishment of the Public–Private Infrastructure Advisory Facility (PPIAF) to help develop strategic frameworks to improve the quality of infrastructure in developing and transition economies. The PPIAF was launched in 1999 as a joint initiative of the governments of Japan and United Kingdom in association with the World Bank. It is a multi-donor technical assistance facility specifically focussing on improving the level of infrastructure services through private sector involvement. PPIAF is governed by a Council consisting of participating donors, which includes bilateral agencies from developed countries, multilateral development agencies and international financial institutions. The Council is supported by an independent Technical Advisory Panel, consisting of international experts in possession of knowledge concerning various aspects of private sector involvement in infrastructure. The objectives of PPIAF are fulfilled through two main channels:

- Providing technical assistance to develop infrastructure strategies to fully engage the private sector.

- Identifying and disseminating best practices of private sector involvement in the delivery of infrastructure services.

PPIAF help to facilitate the transformation of governments from financier, owner and operators of infrastructure services to regulators of privately provide services, and cover a broad range of institutional approaches ranging from management contracts, leases to concessions and divestitures. They provide technical support to facilitate private sector involvement in a range of activities from financing, rehabilitation, operation and management of services. PPIAF funds a range of activities in developing and the transition economies of Africa, Asia, Eastern Europe, Middle East, Latin America and Caribbean. Their activities are also crucial in providing support to post-conflict economies such as Afghanistan, Angola, Kosovo and Rwanda. However, their involvement is generally limited to certain types of technical infrastructure – roads, ports, airports, railways, electricity, telecommunications, solid waste, water and sewerage, and gas transmission and distribution. Application for PPIAF support for country-specific activity generally requires approval from the beneficiary government or governments where a project involves more than one country. Proposals are assessed across a range of criteria consistent with the remit and mission of the Agency and are expected to identify short- medium- and long-term goals to enable progress to be measured against the intended objectives. They fund studies to help countries develop a strategic framework and action plan for governments and donors to determine reform and investment priorities. A key output from the studies or activities is the Country Framework Report, which outlines and describes key aspects relating to:

- Status and performance of key infrastructure sectors
- Policy, regulatory and institutional environment for involving private sector
- Assistance to policy-makers in developing reform and development strategies
- Assistance to private investors in assessing investment opportunities.

Recent examples of projects funded by PPIAF are illustrated in Table 4.1.

Author's commentary

The PPIAF Agency provides a strategic role in the reform or implementation of infrastructure development. The strategic advice or assistance provided could be at local level, national or regional level i.e. relating to a group of countries. As part of their infrastructure development strategies, they prepare reports at the invitation of a country, involving consultations with stakeholders – governments,

Table 4.1   Sample of projects funded by PPIAF

Country	Project	Description of strategic development activity
Ghana	Private participation in the Road Sector	Developing a regulatory and concession framework to provide guidelines, processes and procedures for private participation in developing and managing road infrastructure and related activities
Tanzania–Zambia	Private participation option study on Tanzania–Zambia Railway Authority	Study exploring options for private participation in the Tanzania–Zambia Railway Authority to identify the most feasible scenario for privatisation
China	Options for private participation in water and electricity in Yunnan Province	Feasibility study on a build-own-operate-transfer (BOOT) project or reverse BOOT project to tap, treat and convey the water from the Umbulan spring so as to increase the water supply for Surabaya (capital of East Java) and surrounding cities
Russian Federation	Universal access to telecommunications-strategy and pilot for Russia	Study on access to telecommunications in underserved and isolated areas of the Russian Federation and on the challenges of expanding access; identifying strategies to promote universal access by using private service providers to invest in, scale up, and roll out service in a pioneering transaction
Serbia and Montenegro	Building regulatory capacity in support of private participation in solid waste management in the city of Belgrade	Developing a solid waste management strategy for Belgrade to promote private participation and build regulatory capacity to handle the responsibilities associated with regulation, long-term planning, administration and contract monitoring
Latin America and Caribbean	Regional initiative to build capacity among water and sewerage regulators	Strengthening the Association of Water and Sanitation Regulatory Agencies of the Americas (ADERASA) founded in 2001, by designing regulatory tools for member countries and developing a business plan
Asia	Pre-feasibility study for an Asia private infrastructure financing facility	Reviewing experience in promoting private infrastructure investment including the development and operation of the Emerging Infrastructure Fund and the Community-Led Infrastructure Finance Facility and the experience and knowledge generated by PPIAF's work in Asia

Source: PPIAF (2004).

international agencies, NGOs, private sector, special interest and consumer groups to identify barriers to private sector involvement. PPIAF has also been instrumental in facilitating policy reform or development of legislation in key areas to facilitate private sector involvement in infrastructure.

The Agency also provides support in other areas such as capacity building, sharing global best practices, consensus building and pioneering activities involving some measure of innovation. These activities could take a variety of forms – seminars, workshops, public awareness campaign, study tours, case studies, lessons learned and training toolkits.

Case study 4.2: The Ministry of Infrastructure, Poland

The Polish Government has recognised the need to improve and develop the nation's physical infrastructure to support the economic and technological development required to compete with other member states in the European Union. In October 2001 the Ministry of Transport and Maritime Management was reformed into the Ministry of Infrastructure and was given a wider remit of activities. These included all forms of land transport, airports, communications, construction and architecture, housing policy and management, spatial management and development support, together with the rehabilitation of cities and state aid in the repayment of housing credits. The structure of the Ministry is shown in Figure 4.11.

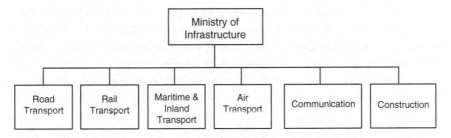

Figure 4.11   Structure of the Polish Ministry of Infrastructure.

A major priority has been given to the development of the domestic road infrastructure network. Funding is being provided by the enactment of legislation on 'paid motorways' in 2003. It is anticipated that the recently formed National Road Fund will provide the means to transform the pace of constructing highway and expressway networks. Furthermore, legal steps have been taken to simplify the buy-out of real estate to make way for new road development. Since the implementation of new legislation in the region of 500 km of new roads and bypasses have been built or are under construction.

The Ministry of Infrastructure is responsible for the state rail network and steps are underway to reform and develop rail infrastructure and to decrease the social cost of functioning rail transport. A restructuring programme accepted by the Polish Government is in process to devolve responsibility for organisation and finance to self-governing regions.

Air transport has been promoted by encouraging low-cost airlines to provide regular services to other European cities. Currently air traffic is expanding fast and a commission has been established to identify a new major airport to serve central Poland.

Policies have been established for the development of the state telecommunications network to incorporate the latest technology with a view to creating an IT informed society alongside the facilities to enable e-business. Emphasis will also be given to improving communication in rural and remote areas.

A programme of housing development is being undertaken based on generally available credit with fixed interest rates allowing for the drawing down of long-term credits. Credit conditions have also been improved by the National Housing Fund and in the first quarter of 2003 housing production increased by 15%.

Author's commentary

In 2004 Poland alongside twelve other east European nations was admitted to the European Union. Great strides have been made since the country was under the control of the Soviet Union in an alliance with other Communist countries in Eastern Europe. However, there is still much to be achieved in developing and upgrading Poland's physical infrastructure in order to allow it to compete effectively with established states in the European Union. This case study demonstrates the intention of the Polish Government to give priority to infrastructure development through the recently formed Ministry of Infrastructure. There are signs that effective national strategies and the devolution of responsibility to regions and local government are beginning to have an impact. Further, there is evidence of progress in the implementation of public–private partnerships and it is encouraging that in the case of road users they are expected to pay for improved services by means of a levy on fuel.

Summary

This chapter has discussed the major considerations required for the development of an infrastructure strategy. The key elements of investment, knowledge and resources, planning and regulatory issues are central to the development of any strategic framework. The need for a holistic approach to understand the complexity of the infrastructure development process and to identify and address systemic failures in infrastructure delivery is critical to the evolution of the strategic framework. It is argued that such an approach is more informed and robust compared to traditional incremental planning. Various problem structuring or systems diagramming techniques are used to illustrate the principles of planning based on a systemic philosophy and to capture a range of infrastructure development issues and processes. The importance of the infrastructure service delivery chain underpinned by the concept of value added and the infrastructure decision-making hierarchy are discussed as crucial in facilitating the development and continuous improvement of infrastructure services. The chapter concludes with case studies illustrating how capacity in strategy formulation could be developed or strengthened. The first case, the PPIAF, demonstrate that a variety of support could be provided externally for developing and transition economies to develop strategic planning capacity by capturing the global experience, knowledge and lessons learned in the development of infrastructure strategies. The Polish Case Study illustrates some of the important principles in developing a national strategy for infrastructure by major restructuring and the creation of a specific Ministry for Infrastructure to develop and co-ordinate strategy.

References

Ahluwalia, M.S. (1997), 'Financing private infrastructure: lessons from India'. In H. Kohli, A. Mody and M. Walton (eds.), Choices for Efficient Private Provision of Infrastructure in East Asia, World Bank, Washington, 85–104.

Brown, D. and Wolfe, J.M. (1997), 'Adjusting planning frameworks to meet changing needs in post colonial countries: the example of Belize', Habitat International, 21(1), 51–63.

Checkland, P.B. (1991), 'From optimising to learning: a development of systems thinking for the 1990's', In R.L. Flood and M.C. Jackson (eds.), Critical Systems Thinking, Wiley, Chichester, pp. 59–75.

Esfahani, H.S. and Ramirez, M.T. (2003), 'Institutions, infrastructure and economic growth', Journal of Development Economics, 70, 443–477.

Fairclough, J. (2002), 'Rethinking construction innovation and research: a review of government R & D policies and practices', Department for Transport and Local Government Regions (DTLR), London, UK.

Fay, M. and Yepes, T. (2003), 'Investing in infrastructure: what is needed from 2000 to 2010', World Bank Policy Research Working Paper 3102, July, World Bank, Washington, D.C.

Flood, R.L. (1995), 'Total systems intervention (TSI): a reconstitution', Journal of Operational Research Society, 46, 174–191.

Hillebrandt, P.M. and Meikle, J. (1985), '*Resource planning for construction*' Construction Management and Economics, 3, 249–263.

Howes, R. (1996), '*Critical systems approach to construction project management*', CIB International Conference, Beijing, China.

Howes, R. and Robinson, H. (2001), '*Urban regeneration activities and the implications for the local construction supply chain*', Proceedings of the RICS Foundation Construction and Building Research Conference (COBRA), In J. Kelly and K. Hunter (eds.), Glasgow, UK, 3–5 September, pp. 375–384.

Kalaitzidakis, P. and Kalyvitis, S. (2004), '*On the macroeconomic implications of maintenance in public capital*', Journal of Public Economics, 88, 695–712.

Kessides, I.N. (2004), '*Reforming infrastructure: privatisation, regulation and competition*', A World Bank Policy Research Report, World Bank, Washington, D.C.

Miller, J.B. (2000), '*Principles and practice of public and private infrastructure delivery*', Academic Publications 2000, Kluwer.

Morah, E.U. (1996), '*Obstacles to optimal policy implementation in developing countries*', Third World Planning Review, Volume 18(1), 79–105.

Pean, L. and Watson, P. (1993), '*Promotion of small-scale enterprises in senegal's building and construction sector: the "AGETIP" experience*' In New Directions in Donor Assistance to Microenterprises, Organisation for Economic Co-operation and Development, Paris.

PPIAF (Public–Private Infrastructure Advisory Facility) (2004), online at http://www.ppiaf.org

Raftery, J., Pasadilla, B., Chiang, Y.H., Hui, C.M. and Tang, B. (1998), '*Globalization and construction industry development: implications of recent developments in the construction sector in Asia*', Construction Management and Economics, 16, 729–737.

Robinson, H.S. (2000), '*A critical systems approach to infrastructure investment and resource management in developing countries: the InfORMED approach*', Unpublished PhD thesis, South Bank University, London.

Wells, J. (2001), '*Construction and capital formation in less developed economies: unravelling the informal sector in an African City*', Construction Management and Economics, 19, 267–274.

Yeo, K.T. (1995), '*Planning and learning in major infrastructure development: systems perspectives*', International Journal of Project Management, 13(5), 287–293.

Part 2

Implementation issues

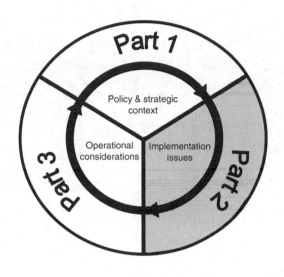

Key themes

project evaluation
procurement choices
finance & funding

Chapter 5

Project evaluation
and resources

Introduction

A variety of factors influence the selection of infrastructure projects. The
contextual or macro variables such as economic, social, institutional and
environmental factors reflecting the policy priorities of a country are
crucial. So too are the project-specific or micro variables relating to the
type of development, design choice, construction methods, location and
duration of the project. Economic growth is a key factor influencing infrastructure investment, but in itself, it is an inadequate criterion for project
selection as development is far more complex than that. Other important
issues must be considered such as socio-cultural, environmental, institutional and political factors. Typically, social development initiatives on
increasing access to education and health services for enhancing human
development capacity are vital in supporting the productive sector necessary to facilitate economic growth. However, an economic programme
involving the construction of large-scale factories and business parks in a
particular region could put small producers and retailers out of business,
affecting the livelihoods of local communities and disadvantaged individuals as unintended consequences. Social development programmes aimed
at improving water supply and sanitation could increase the standard of living and at the same time alleviate serious environmental problems such
as pollution, contamination and the transmission of waterborne diseases.
Economic programmes to create rural employment opportunities have
the potential to improve social status and empowerment, through earning
higher incomes and wages. The above examples illustrate the need to take
account of linkages between different contextual variables, as they are
sometimes negatively or positively interrelated, in the evaluation of infrastructure projects. Project-specific or micro variables reflecting the type
of development, design choice, construction methods, development size,
duration and specific location of projects arc also rclatcd to the contextual
factors. These tend to have a more direct impact on employment, wages,

local communities, standard of living and the environment. It is the inter-relationship, interaction and impact of these contextual (macro) and project-specific (micro) variables that form the basis for developing robust criteria for the evaluation of infrastructure projects. The purpose of evaluation is to examine whether a project or programme is worthwhile from the perspectives of the project owner and where appropriate the national context with respect to certain priorities. Two levels of evaluation are often required. Macro evaluation at the national level by government planners, policy and decision makers focusing on the overall goal of a project in terms of its broad impact on economic, socio-cultural, institutional, political and environmental indicators. Micro evaluation by the project owner or sponsor will be concerned with issues relating to the project's viability within the context of specific objectives. The project owner (whether public or private) is responsible for the implementation of the project, its use and maintenance, and at the national level, the government is responsible (whether local or central government) for the political framework within which the project will evolve. A holistic or systemic view is therefore required in evaluating a project to encapsulate both project- and national level considerations.

Following this introduction, the key contextual (macro) variables such as economic, social, environmental and institutional factors are discussed. Project specific (micro) variables such as design choice, construction methods, size and type of infrastructure development are also examined. The resource implications of infrastructure projects are analysed, and the need for the management of resources to improve the delivery of sustainable infrastructure assets and services is discussed. An outline is provided of the underlying principles, strengths and weaknesses of different types of resource management models. Finally, the InfORMED (**Inf**rastructure **O**rganisation, **R**esource **M**anagement, **E**valuation and **D**evelopment) approach underpinned by systemic principles and incorporating both contextual (macro) and project-specific (micro variables) is presented to facilitate project selection and resource management.

Macro or key contextual variables

There are several fields of appraisal but priorities are continuously evolving depending on a country's policies. Significant changes have occurred in appraisal methodology over the years, some of which are more recent due to the sustainability debate. Increased emphasis is now being given to environmental considerations to reflect the concern of global warming, particularly by some international development agencies and industrialised nations. Project owners or sponsors are therefore gradually adopting more rigorous techniques to assess the environmental impact and sustainability of infrastructure projects. Institutional appraisal has become an increasingly important dimension, as the outcome of projects depends on the quality of institutions responsible for implementation and operation. The financial aspects of a project have always been a crucial factor but emphasis has been too much on initial capital costs during appraisal stage at the expense of recurrent financing or project life cycle costs. This practice is increasingly recognised as inadequate leading to unsustainable infrastructure assets. Recent procurement practices such as public–private partnerships (PPP) and the UK private finance initiative (PFI) have brought the concept of life cycle cost into sharper focus as the responsibility for design, construction and maintenance are integrated, and the operational risk is shifted from the public to the private sector. The importance of key appraisal fields are discussed as follows.

Economic considerations

The linkage between infrastructure investment and economic development is complex. Economic theory suggests that infrastructure investment enhances the productivity of both capital and labour in the production process. Adequate infrastructure services therefore lowers production costs and increases profitability. Various studies have established a positive and significant relationship between the levels of infrastructure investment and economic growth. Other studies have also shown that public infrastructure capital has a significant effect on production decisions at a regional level. However, the precise relationship defined as the causality running from infrastructure investment to economic growth is still a matter of intense debate among economists. Adequate infrastructure is also a key factor that influences development investment and the level of competition in international trade. A study by Amjadi and Yeats (1995) noted that the high transport costs caused by failure to maintain or improve ports and other transport infrastructure have contributed to the relative decline of exports in some developing countries.

There are short- and long-term effects from infrastructure investment. First, infrastructure investment is often a 'public works' activity directly influencing personal income in the short term through employment and wages in the construction sector. Then there is also a multiplier effect in the local economy since the marginal propensity to spend increases creating further employment and wages in other sectors. Second, personal income, in part, determines the level of savings and willingness to pay for infrastructure services, which in turn generates further investment. Third, infrastructure investment is an input into the production process; hence there are productivity benefits generated through the use of infrastructure assets in the long term. Investments or accumulated investments (capital stock) influence personal income through its effects on the marginal product of labour. However, the extent of the gains in the local economy crucially depends on the type of infrastructure and the capacity of local construction and other sectors.

81

Socio-cultural considerations

Access to infrastructure services is fundamental in defining welfare and poverty. Infrastructure affects the dimensions of poverty in terms of employment opportunities and income creation, which determines the standard of living. Studies have shown that the availability of transport infrastructure in the periphery of urban areas is a key factor in obtaining employment. A lower standard of living as a result of limited or lack of employment and income opportunities creates enormous problems such as crime, vandalism, homelessness, drug culture and other anti-social behaviour. These problems are exacerbated in regions experiencing rapid increases in population and urbanisation creating further pressure on available infrastructure. A key consideration in providing infrastructure is therefore to protect communities, improve social status through creating employment opportunities for the disadvantaged and poor communities, to improve their ability to pay for services. Infrastructure programmes (e.g. parks, recreational and sports facilities) help in revitalising urban areas, and serve to galvanise individuals and local communities.

The construction of infrastructure contributes to poverty reduction through short- and long-term employment opportunities as discussed in the previous section. However, the level of social impact and its sustainability depends on the nature and type of infrastructure programmes, as some are more employment intensive than others during construction and post-construction phases. Infrastructure alone does not create employment and income in the post-construction phase in regions without development potential. It is therefore important that infrastructure programmes are implemented alongside other initiatives to enhance development potential, so that employment opportunities created are sustained during the operation of the assets.

Environmental considerations

Infrastructure facilities create various economic and social benefits to humankind but they raise different and at times complex issues concerning the natural environment. Environmental considerations are an integral part of project evaluation, and should not be added on as afterthoughts. Environmental impact statements are becoming increasingly important, and in some internationally funded projects, they are mandatory in project evaluation. Clean water and sanitation, electricity services, safe disposal of solid waste, better roads, better education, health, recreational facilities and services provide various environmental benefits.

Expansion of road networks mitigates pollution associated with traffic congestion. However, road schemes, if not properly planned, designed, constructed and managed could lead to significant deterioration of the natural habitat resulting in increased air, noise and water pollution and poor drainage and sanitary conditions. These adverse conditions particularly in crowded neighbourhoods and communities will increase the risk of diseases and their transmission. The continuous pressure or demand to improve the level of infrastructure provision, as a result of such adverse conditions exacerbated by population growth and urbanisation, poses a major problem for the environment.

Institutional considerations

The purpose of institutional appraisal is to identify existing capability and the need for measures to enhance institutional effectiveness. The view that infrastructure provision should be dominated by the public sector is now being questioned due to poor or deteriorating performance of public sector organisations. Institutional changes are desirable to promote efficiency, fairness, accountability in the supply of infrastructure services, and to attract large investments from the private sector. National governments and international development agencies are rethinking the concept of infrastructure provision. Recent development in PPP such as private participation in infrastructure (PPI) and PFI have shown that the private sector is increasingly recognised as an important actor in the delivery of key public services in health, education, transport and other sectors. The gradual involvement of the private sector reflects the need for the application of commercial principles, encouraging and broadening of competition and the increased response to users' needs. Various types of public–private partnerships exist to attract investment in infrastructure, increase efficiency and to expand infrastructure services. However, the level of private sector participation and impact depends on the policy and strategic framework and the capacity of governments to plan and co-ordinate investments, monitor and regulate infrastructure delivery.

Political considerations

This is perhaps the most crucial factor in any type of infrastructure development as governance, stability and human rights issues have become important to investors. Political conditions influence whether a project is likely to go ahead and has far reaching implications for projects in situations of high risk and uncertainty. The drive to implement PFI in the UK is not only to help improve public services in key sectors of health, education and transport, but it also influences voting outcomes as infrastructure projects have a high visibility in local communities. Developed nations tend to be characterised by stable democratic structures that are capable sustaining infrastructure development programmes and changes in governments. This contrasts sharply with many developing countries with poor democratic or governance structure, human rights abuse and civil wars that have a significant negative impact on the interest of investors or donors in funding infrastructure projects. It is therefore important to assess the political situation since projects that are disrupted or have to come to a standstill are costly in terms of cost, time, deterioration of uncompleted infrastructure and wasted resources. Countries affected by war and civil conflicts, or those emerging from it, have the greatest need for infrastructure but pose a major challenge for local and international private investors in terms of risk. First, there is usually a political vacuum resulting in lack of clear policy priorities and strategies, and an inadequate government capacity to implement new projects leading to massive disruption in infrastructure services to support development. Second, transactions involving public budgets, where they exist, are at best opaque and often characterised by a climate of unaccountability. In extreme cases, limited funds available to support development are used to fund highly visible, politically motivated infrastructure projects often becoming white elephants, or diverted to procure military hardware and installations for regime protection and civil war activities.

Micro or project-specific variables

There are number of variables that impact on the outcome of a project. These 'micro' or project-specific factors are development type, design option, construction methods, implementation period and the size of development. Project-specific variables are of critical importance in evaluation as they affect project costs and benefits, the resource requirements, project viability and associated income and employment opportunities. For example, projects requiring labour-intensive technologies are normally executed by small construction firms, creating employment for local SMEs, subcontractors, both in the skilled and unskilled labour markets. These types of projects cannot be dominated by large organisations given the labour intensity and the site-based method of production that provides smaller firms with some advantages. On the other hand, large and complex projects will almost certainly require the services of large national or international firms with a wealth of hi-tech equipment and plant for construction, and a highly skilled workforce with technical and management expertise. Changes in project variables therefore have significant implications on opportunities for local consulting and contracting organisations, employment creation, income and wages, social and skills development, environmental impact, which in turn affects the contextual (macro) variables such as economic, social, institutional and environmental factors.

Development type

The development type reflects whether projects/programmes are new construction, rehabilitation works or mixed development. Rehabilitation works are needed to prolong the economic and physical life of infrastructure assets and to enhance their functionality by updating standards and incorporating modern technology. However, they tend to be difficult and expensive because of the risk, uncertainty and constraints such as noise, unsociable or restriction on working hours and the normal usage of existing services associated with its development. In developed countries, the proportion of rehabilitation or renewal works are significantly higher compared to developing countries because of the massive stock of built infrastructure assets.

Design issues

Design reflects the form, layout or structure of infrastructure and its associated aesthetic features. There are also issues regarding choice of materials, products and finishes closely associated with the form and structure. Design options range from traditional design using the most basic local materials (as in rural areas of developing countries) such as mud, bamboo and thatch/grass to innovative design such as very tall buildings and sophisticated bridges like, the London Millennium Bridge using modern steel and concrete materials. Innovative design typically requires materials such as prefabricated building components, hi-tech lifts, curtain walling, concrete beams and steel frame structure, intelligent services and building fabric. Design options for roads range from earth roads to surfaced roads with high-quality materials for subgrade, subbase, base and wearing courses.

Construction issues

Construction issues relate to the management and technical processes used for implementing infrastructure projects. The management and technical processes depend on the type of construction. Standard designs are effectively managed by programmed organisations relying heavily on routine and standardised construction procedures. Innovative designs require highly flexible management structures to control complex construction processes. Technology choices are also influenced by design. Technical processes range from labour-intensive approaches to equipment (automated) processes relying on plant, machinery or robots for on-site construction or the assembly of prefabricated materials. Labour-intensive approaches are suitable for simple construction relying on semi-skilled/unskilled labour using basic hand tools and light equipment. Equipment-intensive approaches are used in advanced construction relying on a small highly skilled and productive labour using heavy plant and machinery. Table 5.1 shows the technological options for various construction activities.

Development size

The development size generally reflects the level of capital investment required for planning, designing, constructing and commissioning infrastructure projects. However, it is increasingly recognised that there is a strong relationship between capital costs and life cycle costs required for the operations and maintenance of completed infrastructure projects. There is also a growing awareness of the increasing cost of advanced technology in modern infrastructure. This is as a result of the significant impact of technology in the way governments, businesses, people and society operate.

Table 5.1   Technological options for various construction activities

Construction activity	Labour-intensive	Intermediate	Capital-intensive
Excavate soil	Hoe, shovel, pick	Excavator	Mechanical excavator, dozer
Excavate rock	Pick, crowbar	Hand drill/blast	Compressed air drill/blast
Load	Hoe	Shovel	Mechanical excavator
Haul	Headbasket/ wheelbarrow	Animal cart/tractor	Tipper truck
Spreading	Hoe, rake	Hoe, rake	Dozer, grader
Compacting	Hammer, rammer	Animal towed roller Hand propelled roller	Tractor towed roller Self-propelled roller
Haul/lay bitumen	Stretcher/ rake	Wheelbarrow/ hand propelled screed	Truck/paver
Mixing concrete	Small drum mixer	Medium drum mixer	Large mixer
Hauling/ placing concrete	Wheelbarrow	Hoist/concrete pump	Crane/skip

Different technologies are incorporated in modern buildings and facilities depending on the type of design or infrastructure sector (e.g. traffic management systems, security systems, solar or renewable technologies, health technologies, telecommunications, air-conditioning, heating, water supply, waste water and power systems). The implication is that the cost of the traditional 'shell and core' or fabric is gradually decreasing compared to the technology components in infrastructure. There are also differences and disparities in the application of technologies in developed and developing countries with significant impact on the level of development investments.

Implementation issues

The implementation period is the time set to realise project or programme objectives within a given planning horizon. The preparation and implementation of infrastructure projects can take a long time, often several years. There are various time-scales associated with the different phases (planning, design and construction) of infrastructure development, which varies depending on the type of infrastructure and the country. The time-scales are influenced by a number of factors including client needs, country context such as resources available, planning laws; location of facility whether in rural, urban areas, greenfield or brownfield sites; project context such as type of consultants and contractors involved, scale of development, procurement options, design and construction options.

Resource implications of infrastructure projects

Infrastructure projects require resources at various stages of development for identifying and screening projects, developing and structuring them for financing, reviewing planning applications, carrying out regulatory and enforcement activities during service delivery. Other resources such as design labour, construction labour, plant and materials are directly involved in the physical transformation of projects into infrastructure, and in the operation and maintenance of completed infrastructure facilities. Figure 5.1 shows the inputs required for the production of infrastructure facilities.

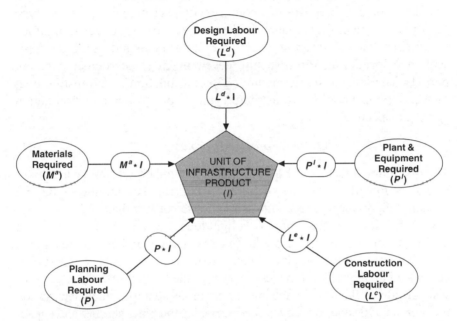

Figure 5.1   Inputs required for infrastructure production.

Two types of entity sets can be distinguished in the diagram. Elementary entity sets are single element sets such as infrastructure (I), for example, roads, power stations, schools, hospitals, parks, factories and offices, etc. Similarly, construction labour and materials are elementary entity sets. Examples include steelworkers, carpenters, electricians, painters, bricklayers, masons, plumbers, etc.; and for materials, aggregates, glass, cement, pipes, steel reinforcement, timber and other products. Elementary entity sets interact to form complex entity sets. For example, construction labour (L^c) and infrastructure (I) interact to form a complex entity set ($L^c * I$) – the construction labour production coefficient. This is the construction labour requirement for the production of a unit of infrastructure. A 'unit of infrastructure' is a conceptual term expressed in various physical quantities (area, number of users, or some other measures of output). Building-type infrastructure such as schools, hospitals, police stations and houses are usually measured using the superficial area method (gross floor area).

87

There is also the unit method based on function such as 'workspaces' for offices and factories; 'bed spaces' for houses, hotels and hospitals; 'pupil places' for schools and colleges; and 'seats' for churches, theatres and community centres. However, there is no universal method for non-building type infrastructure. Non-building type infrastructure, mainly civil engineering structures (e.g. roads, railways, sewerage and ports) are measured in a variety of ways – superficial area, cubic volume, linear (length or width of facility), number of users or other unit of output measures.

Level of infrastructure provision and production

The development objectives of a country will determine the scale and type of infrastructure projects required. These objectives are reflected in policy instruments such as structure, local and master plans indicating the type, scale and location of infrastructure development for particular planning periods. The level of infrastructure (IQ_z) is a function of the institutional factors (IF) supporting infrastructure provision (*prov*) and production (*prod*), respectively:

$$IQ_z = f(IFprov, \; z\,IFprod, z) \quad \forall z \tag{1}$$

The institutional factors in Eq. (1) reflect the number, type and capacity of providers for different types of infrastructure (z). The number, type and capacity of providers for each infrastructure sector is influenced by the regulatory structure. A 'free entry' or a light touch regulatory structure in the health sector, for example, could generate large numbers of healthcare providers in the private sector. A heavy regulatory regime in water supply could lead to only a few providers. However, the level of infrastructure provision is also influenced by the implementation capacity, which depends on the number, type and capacity of producers, i.e. the type, number and capacity of manufacturing, design and construction organisations, and increasingly in developed countries, specialist facilities management firms.

Resource implications and management

The quantity of resources required or demand (D) in Eq. (2) is a function of the quantity of infrastructure to be produced (IQ_z) and the production coefficients (C), i.e. the resource requirements for a unit of infrastructure produced:

$$D = f(IQ_z, C) \tag{2}$$

The production coefficients (C) specific to each type of resource vary according to the type of project, development type, design and construction technology options. Similarly, in Eq. (3), the level of resources available or supply (S) is a function of existing level of resources (E), resource development growth (G) and productivity rates (δ):

$$S = f(\delta, E, G) \tag{3}$$

Every type of infrastructure requires planning resources, i.e. for reviewing projects, structuring them and arranging for finance, carrying out regulatory and enforcement activities. The rate of resource (materials, design labour, construction labour and plant) consumption during infrastructure production depends on the type of infrastructure, and the production coefficients. The quantity of planning, design labour, construction labour, materials and plant resources required are illustrated in Eqs. (4)–(8):

$$RQ_j = \sum_z C_R * IQ_z \qquad (4)$$

In Eq. (4), there are j types of planning labour resources (R). The types of planning resources could be, for example, town planners, building control officers/planners, building inspectors, health and safety inspectors, environmental inspectors, enforcement officers and regulators. Similarly, in Eq. (5) there are n types of design labour (L^d), such as architects, surveyors, and various types of engineers (civil, aerodrome, transport engineers, water and sanitation, building services, electrical and power, etc.).

$$L^d Q_n = \sum_z C_{L^d} * IQ_z \qquad (5)$$

Equation (6) also shows p types of construction labour (L^c), while Eq. (7) shows m types of components or material resources (M^a). There are many construction materials and components in the market with significant implications for resource management. Sir John Egan (Egan, 1998) in his review of the UK construction industry titled 'Rethinking Construction' noted that a house has about 40 000 components compared to 3000 for an average car. He cited the example of about 150 different types of toilet pans in the UK compared to six in the USA and argued for clients and designers in the UK to make much greater use of standardisation to improve efficiency:

$$L^c Q_p = \sum_z C_{L^c} * IQ_z \qquad (6)$$

$$M^a Q_m = \sum_z C_{M^a} * IQ_z \qquad (7)$$

Equation (8) shows o types of equipment and plant resources (P). While developed countries have different types or a vast range of equipment and plant resources, the types of equipment and plant resources are often limited and basic in many developing countries:

$$PQ_o = \sum_z C_P * IQ_z \qquad (8)$$

The availability of the different types of resources outlined above depends on the existing level of resources (E), resource growth rates (G) and the productivity (δ). The quantity of planning, design labour, construction labour resources, materials and plant resources available are illustrated in the following equations.

$$RS_j = \delta\{E_j + (E_j * G_j)\} \qquad (9)$$

$$L^d S_n = \delta\{E_n + (E_n * G_n)\} \tag{10}$$

$$L^c S_p = \delta\{E_p + (E_p * G_p)\} \tag{11}$$

$$M^a S_m = \delta\{E_m + (E_m * G_m)\} \tag{12}$$

$$P S_o = \delta\{E_o + (E_o * G_o)\} \tag{13}$$

Infrastructure resource–cost relationship

Infrastructure resources and costs are intrinsically linked. A scarcity of resources means that unit resource costs are likely to increase leading to an overall increase in infrastructure development costs. Infrastructure development costs are therefore affected by the demand and supply situation in the resource markets. The total infrastructure cost is likely to be the sum of resources consumed at all development stages. Capital costs are incurred from project proposal to construction stage, and recurrent (life cycle) costs are incurred during operation and maintenance or service delivery. The costs could be determined for each stage and type of infrastructure based on resource production coefficients, resource consumption and unit resource costs. For example, planning costs are associated with reviewing planning applications, carrying out regulatory and enforcement activities during infrastructure service delivery. This cost is incurred in the form of planning consultants' fees, planning application fees, building regulations and 'planning supervisors' fees. In some developing countries, planning fees are very low and do not always reflect the full cost of evaluating planning applications. As a result, planning, regulatory and enforcement activities are often compromised, increasing the tendency for corruption and the delivery of substandard infrastructure facilities.

Design costs are in the form of professional fees charged by architects, engineers, surveyors and other specialists. This is usually based on a percentage of the development cost or recommended scale charges for each profession. The fee levels could also be time dependent reflecting the complexity and scale of particular types of development. The total construction cost is the sum of resources utilised in the construction stage (materials, construction labour and plant) plus an organisation's management/supervision cost (overheads). Infrastructure projects are profitable if benefits or expected revenues exceed the total development costs incurred.

Traditional cost structure of infrastructure projects is normally presented in the form of static elemental cost plan. While this approach provides estimates of likely infrastructure investment requirements, there are obvious limitations to its use for resource management. The alternative resource-based planning approach provides not only estimates of cost requirements, but more importantly a better understanding of the resource mix and the implication for infrastructure development. This will enable changes in the types of labour, material and plant resources required to be carefully managed against the supply of these resources.

90

The challenge for development planners is therefore to ensure that an equilibrium or near-equilibrium state is achieved so that resources available or supply (S) is balanced against resources required or demanded (D). Failure to achieve this would have serious consequences for development resulting in different types of failures. Type I failure illustrates a situation where resource requirement is lower than the available supply (resource pool limit). Type II failure is the reverse situation where resource requirement (e.g. cement) is greater than the available supply. There is also a third type of failure (Type III), which is structural in nature reflecting a mismatch between the structure of demand (project profiles) and that of supply, i.e. number, type and capacity of producers (see Figure 5.2).

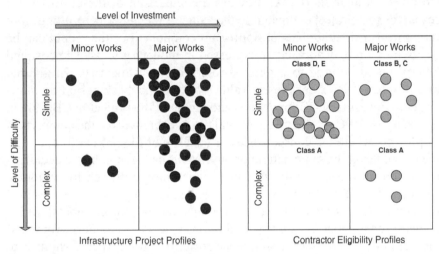

Figure 5.2   Example of type III failure.

In the context of the structural failure shown in Figure 5.2, the problem could be addressed by looking at a combination of demand and supply side strategies. Demand measures could include repackaging projects by reviewing project variables such as development size, design and technology options to reflect local capability and capacity constraints. Supply measures, for example, could include improving local manufacturing capacity, strengthening design and construction capacity by encouraging mergers and acquisitions of construction firms, joint-venture with foreign firms aimed at developing expertise in key areas. Developing resources or capacity building is crucial for infrastructure development. Capacity building adds value to infrastructure development through the acquisition of the necessary managerial and technical skills, new techniques and technologies to enhance the infrastructure delivery chain. Adequate planning resources during the development process adds value by ensuring that satisfactory standards are set and maintained, regulations are fully enforced using

91

appropriate building and development control legislation and projects are implemented in accordance with structure, local and master plans. A lack of resources undermines the value added to the infrastructure delivery chain. Low planning and regulatory capacity therefore slows development and implementation with significant impact on the speed and efficiency of the infrastructure delivery chain.

Infrastructure evaluation models

Traditional project evaluation techniques such as cost benefit analysis, cost-effectiveness, and utility analysis focus on costs and the likely benefits. Cost–benefit analysis (CBA) provides a comparison of the cost of input resources to a project compared to the value of the benefits resulting from the project. It is used when a significant component of the output can be easily measured, quantified or expressed in monetary units. The cost and benefit could be used to determine return on investment (ROI), internal rate of return (IRR), net present value (NPV) or payback period on infrastructure investments. Both outputs and inputs of the project in CBA are in monetary units. Cost effectiveness analysis (CEA) involves the comparison of projects where the consequences or benefits of the project (output) are measured using the same natural or physical units e.g. number of accidents prevented for transport infrastructure, reduction in pollution for factories. The output in CEA may be in any units, while the input is in monetary units. Cost utility analysis (CUA) involves a comparison of project (inputs), which is measured in monetary units with the consequences, or benefits (outputs) measured using utility or a preference scale. It is used when a significant component of the project output *cannot* be easily measured, quantified or expressed in monetary units. The output in CUA is in dimensionless (utility) units reflecting some measures of satisfaction while the input is monetary units. Alternative multi-criteria evaluation methods such as simple non-compensatory, additive weighting methods, analytic hierarchy process (AHP) are also useful to complement these traditional techniques as they capture a range of other non-economic criteria (environmental, social, institutional, political factors) not easily expressed in monetary units (Rogers, 2001).

More significantly, there is a need for resource evaluation to be carried out alongside these traditional, or multi-criteria evaluation techniques, as infrastructure resources and costs are intrinsically linked. The importance of a resource-based approach to understanding infrastructure development is highlighted in the following quote from Morton and Jaggar (1995): '*if we wish to understand why a building costs a certain amount, we need to know more about the resources which go into it*'. There are different types of infrastructure resource models that could be used to understand and facilitate the infrastructure development and management process. The underlying principles, strengths and weaknesses of various models are outlined as follows:

Turin/Strassman model

Macroeconomic approaches such as the Turin/Strassman paradigm (Strassman, 1970; Turin, 1978) are seen as aggregate resource planning tools, as the estimation of global construction output is one of the main components. A fundamental criticism of the Turin/Strassman paradigm is its implicit assumption that resources can be efficiently allocated inter-sectorally and between alternative construction outputs due to the existence of market forces (Drewer, 1997). The notion of construction capacity is meaningless unless it is disaggregated into some kind of input–output matrix relating to different levels of design and technology on the demand side, and also to different resources on the supply side. Further insight can only be gained by breaking down capacity in terms of its elements – project types, design and technology, contractors, consultant types and associated labour, materials, plant and equipment resources.

Miller and Evje's model

Miller and Evje (1999) developed a model called CHOICES for infrastructure project selection using procurement or project delivery methods as variables. The approach optimises infrastructure portfolio to meet the strategic goals of clients within capital constraints in each year for the duration of the analysis. However, project evaluation methods cannot ignore the impact of non-financial resource constraints on infrastructure delivery. The approach is very useful and applicable in advanced economies where resources for the implementation of infrastructure projects are generally well developed. However there are limitations in the application of this approach in developing countries, where resources such as materials, equipment, various planning, design and construction inputs crucial for infrastructure development pose equally, if not more, challenging problems.

Resource planning models

Several types of resource planning models have been developed. These include manpower models by Uwakweh and Maloney (1991) and Rosenfeld and Warszawski (1993) for forecasting construction labour resources. Other resource planning approaches include Hillebrandt and Meikle (1985) and Lemessany and Clapp (1978). The underlined objective of resource planning is to be capable of identifying the resource implications of construction projects implied by the development plan. The approach focuses on quantifying resource requirements and is therefore useful in situations where resources are not properly developed and there is a need to define a strategy for the long-term development of indigenous construction capacity.

Limitations of existing resource models

The resource planning approaches outlined above while useful are inadequate for a number of reasons. First, some of the approaches are too static, limited or restrictive in scope, as they do not consider the totality of resources in infrastructure development. Second, they do not adequately address the dynamic interactions between actors and resources in the infrastructure development chain. By means of example, the importance of value-added activities of planning and regulatory functions in infrastructure delivery are ignored. Finally, the policy variables and its interaction with the strategic context are not explicitly or adequately addressed.

The InfORMED approach

InfORMED (**Inf**rastructure **O**rganisation, **R**esource **M**anagement, **E**valuation and **D**evelopment) is a systemic approach encapsulating the dynamic interaction in the infrastructure delivery chain. The model is underpinned by a policy and strategic framework to understand and facilitate infrastructure development (Robinson, 2000).

The policy component

The policy component provides the basis for determining the relative importance of policy macro variables (economic, social, environment, institutional and international factors), and their impact on project investment priorities. Figure 5.3 is a generic policy model showing policy clusters, variables and their horizontal interactions.

The policy component reflects the dynamic soft variables, which could be increased, reduced or their intensities altered to match policy priorities. The AHP technique is used for setting priorities based on pair-wise ratings assigned to policy objectives.

Interactions between policy and strategic levels

The interactions between the policy level and the strategic level provides the basis for making infrastructure development decisions based on contextual (macro) and project specific (micro) variables discussed in previous sections, using the InfORMED approach. This linkage between the policy and strategic variables representing vertical interactions is shown in the brainstorm view, using Criterium DecisionPlus Software, with project variables (alternatives) as shown in Figure 5.4.

Projects (alternatives) are rated for their likely impact on, or contribution to policy objectives. Where contributions to policy objectives are low, the project variables can be reviewed to improve contribution scores. The choice of infrastructure project selection reflects the intensity of the vertical interactions between policy variables and strategic variables. For the purpose of comparison, an example of the project rankings under the three different policy scenarios are shown in Figures 5.5a–c.

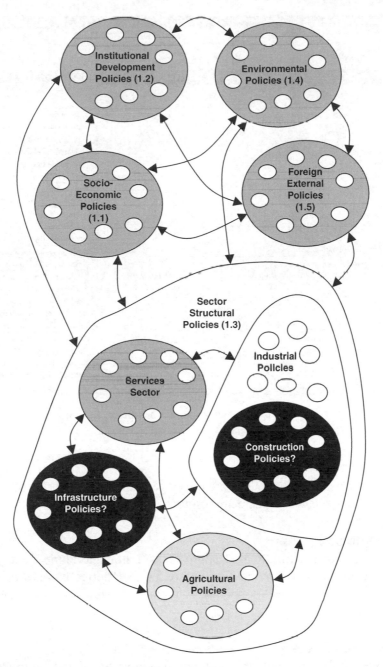

Figure 5.3   Generic Policy Model.

The above results demonstrate the effects of changes in policy variables on strategic options reflected in the ranking on infrastructure projects. The vertical interaction between the strategic and policy level reflects the contribution of infrastructure projects to various policies, the requirement for

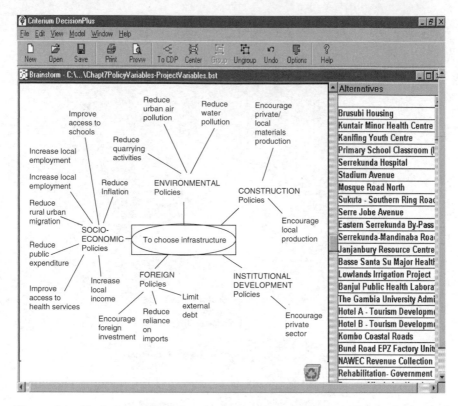

Figure 5.4   Brainstorm view showing interactions between policy variables and strategic project variables (alternatives).

policy intervention where there are policy gaps and/or the ranking of infrastructure projects where there are resource gaps.

Strategic component

At the core of the InfORMED approach is a strategic component – an infrastructure resource model (see Figure 5.6), supported by *hard* project and resource variables, horizontally interacting with *soft* institutional variables.

Built-in parameters

The built-in features include *resource production coefficients, infrastructure cost per unit rates and resource unit prices*. Resource production coefficients give the quantity of resources required per unit of infrastructure output. Technical coefficients (resource requirements per units of kg, m^2, m^3, litres, man-hours, etc. per square metre of infrastructure output or other performance unit) are used, as they are more stable than cost coefficients over time. The built-in features are not constant as such but do vary according to the specific needs of a country.

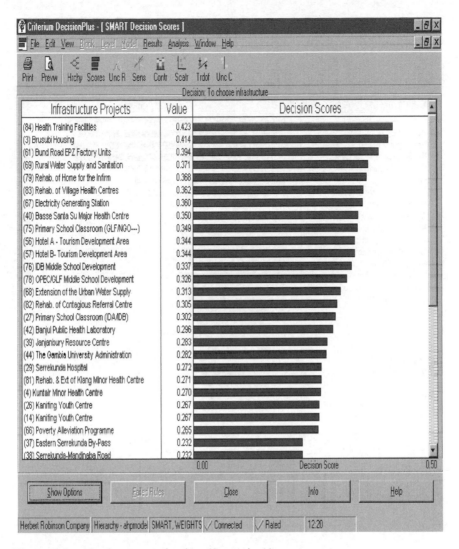

Figure 5.5a   Socio-economic view (Scenario A).

Performance variables (output data)

The model gives information on the following output data: (1) values for the demand of individual resources; (2) values for the supply of individual resources; (3) a resource status index measuring the differences between the demand and supply of individual resources; and (4) infrastructure projects, infrastructure cost breakdown by resources and resource components. Table 5.2 is an example of the performance variables, i.e. the material resource implication of implementing a housing infrastructure project with the following demand subvariables – new development (DevOpt 1), conventional design (DesOpt 2) and a labour-intensive technology (ConsOpt 1).

Table 5.3 shows the likely impact on construction labour resources if all the infrastructure projects/programmes selected are implemented. This

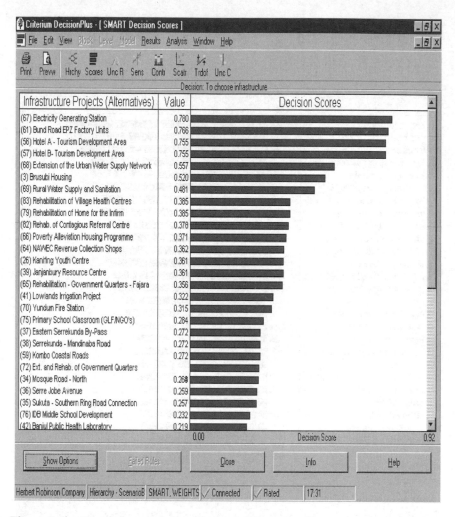

Figure 5.5b   Foreign sector view (Scenario B).

scenario is based on low-resource (supply) growth rate assumption ranging
from 0% to 5% for most of the resources.

Clearly, implementing the entire programme of infrastructure projects
would pose serious resource difficulties. The majority of resources, in the
form of materials, construction labour, design and planning manpower,
show serious deficits and extremely low cover ratios. The resource cover
ratios for the construction labour categories are extremely low, ranging
from 0.61 for formwork carpenter to 0.02 for plant operator. The result
indicates that there is a need for a wide range of capacity building initiatives, or interventions, to address the mismatch between resource demand
and supply. Three types of capacity building interventions can be distinguished: project, resource and institutional intervention. Nevertheless,
the hard project and resource measures alone are not sufficient. A 'soft'

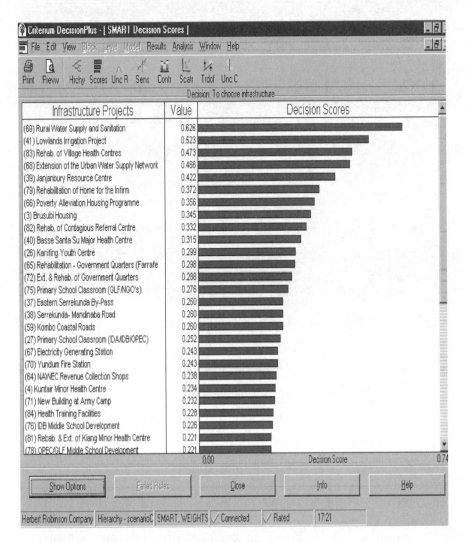

Figure 5.5c   Environment view (Scenario C).

approach is often needed, i.e. institutional changes are required to complement 'hard' project and resource approaches to facilitate infrastructure
development (see Figure 5.7).

In order to achieve the desired performance (output), strategies may
need to be changed, which could mean changes in policy as well. Strategic
project variables could be altered to increase the policy impact of a particular project. To increase the contribution to the policy variable (increase
local employment) design and construction technology could be based on
conventional design and labour-intensive methods.

These 'hard data' reflecting the demand and supply situation together
with the 'soft data' environment therefore provide a coherent framework to
explore strategies for infrastructure development. Resource requirements

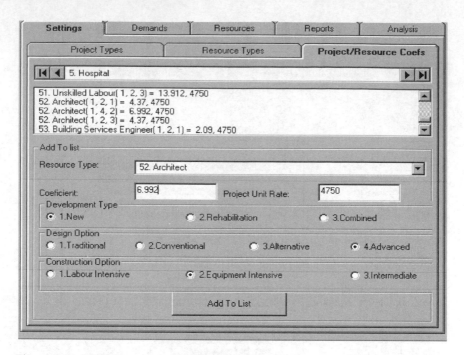

Figure 5.6   Model components showing project/resource coefficient view.

Table 5.2   Material variables for conventional design option

Project ID 3	Project name Brusubi Housing	Infrastructure type Housing	Development size $164,312,500	Project unit rate $2,750	
Res. ID	Resource type	Resource coefficient	Resource quantity	Resource unit price	Resource cost	Resource share
3	Sand	1.00	59750.00	135.00	8066250.00	0.049
4	Aggregate	0.58	34834.25	200.00	6966850.00	0.042
8	Cement	339.65	20294147.25	1.08	21917679.03	0.133
15	Construction timber	0.02	956.00	3750.00	3585000.00	0.022
17	Joinery timber	0.01	358.50	6375.00	2285437.50	0.014
18	Steel reinforcement	14.10	837097.50	7.70	6445650.75	0.039
19	Structural steel	18.11	1081773.75	6.00	6490642.50	0.040
20	Roof sheeting	21.68	1295499.50	15.00	19432492.50	0.118
21	Paint	1.67	99543.50	50.00	4977175.00	0.030
22	Ceiling/ part boards	1.88	112509.25	87.00	9788304.75	0.060
23	Floor tiles	1.57	93807.50	130.00	12194975.00	0.074
24	Wall tiles	0.23	13921.75	110.00	1531392.50	0.009
25	Toilet suites	0.02	932.10	1200.00	1118520.00	0.007
26	Washhand basins	0.02	932.10	1075.00	1002007.50	0.006
28	Baths	0.02	932.10	1575.00	1468057.50	0.009
29	Kitchen sinks	0.02	932.10	1325.00	1235032.50	0.008
31	Pipes	0.11	6393.25	96.00	613752.00	0.004
32	Glass	1.17	70086.75	108.00	7569369.00	0.046
				Total $	116688588.03	0.710

Table 5.3   Construction labour resource output (for 37 projects)

Res. ID	Resource type	Unit	Unit price	Resource req.	Resource availability	Resource status	Cover ratio
48	Plant operator	man hrs	10	758193.45	12075.00	−746118.45	0.0159
49	Tarmac/asphalt layers	man hrs	5.42	19021.08	1288.00	−17733.08	0.0677
35	General foreman	man hrs	10	434285.75	42336.00	−391949.75	0.0975
50	Other skilled labour	man hrs	5.42	79229.00	10500.00	−68729	0.1325
36	General skilled labour	man hrs	5.42	744064.02	101920.00	−642144.02	0.1370
38	Steelfixer	man hrs	5.42	117660.30	16744.00	−100916.3	0.1423
42	Structural steel/ metal worker	man hrs	5.42	309809.47	44520.00	−265289.47	0.1437
44	Tiler	man hrs	5.42	216120.80	30940.00	−185180.8	0.1432
37	Concrete worker	man hrs	5.42	635199.20	101920.00	−533279.2	0.1605
46	Plumber & drainlayer	man hrs	5.42	125829.21	20608.00	−105221.21	0.1638
45	Electrician	man hrs	5.42	263977.52	58240.00	−205737.52	0.2206
41	Joinery carpenter	man hrs	5.42	795384.79	207872.00	−587512.79	0.2613
39	Mason	man hrs	5.42	1044552.53	289886.25	−754666.28	0.2775
43	Painter & decorator	man hrs	5.42	163274.63	48384.00	−114890.63	0.2963
51	Unskilled labour	man hrs	3.61	5504678.79	1645056.07	−3859622.72	0.2988
40	Formwork carpenter	man hrs	5.42	128793.79	78400.00	−50393.79	0.6087

are affected by hard project variables such as design, technology, infrastructure type, investment levels, location and implementation period. The structure of demand also influences the structure of supply, which is determined by the capacity of the production units to provide services in relation to the type, size and locational distribution of the infrastructure projects. Improving the supply situation through the availability of resources, can be influenced by soft institutional variables. For example, the procurement route which affect resources is not simply determined by the 'project holder organisation', but is often influenced by 'investment organisations' or funders. Design and technology choice may be indirectly influenced by 'planning and regulatory organisations'. Similarly, the project type is not solely determined by 'project holder organisations' but is strongly influenced by 'investment organisations' providing funding. Soft institutional variables can alter the way in which infrastructure projects are procured, funded, designed and constructed in accordance with planning and regulatory standards. The structure of demand may be changed through procurement practices that limit the size of tender packages to certain levels as in the UK PFI procurement. The treatment of these variables as 'soft' is not only consistent with the lack of empirical knowledge about the exact quantitative relationships, but it is a crucial recognition that their influences on hard project and resource variables. In fact, a model that recognises these

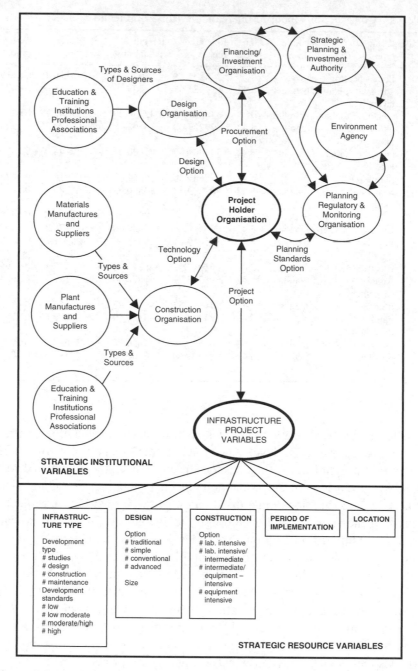

Figure 5.7 Relationship between strategic soft institutional and hard project/resource variables.

variables, although soft, is far superior to those models that include only a limited number of quantitative variables.

The InfORMED model has the potential to provide a basis for making informed decisions at the contextual policy (macro) and strategic (micro) levels and assessing direct impact on the utilisation of available resources, either indigenous or foreign. The model therefore has the ability to assist in establishing priorities and setting infrastructure budgets according to the availability of resources and what can be afforded.

Case studies

Case study 5.1: The Public Works Agency Model (AGETIP)

A public works and employment agency called AGETIP, a French acronym for Agence d'execution des travaux d'interet public contre le sous-employ is a dual agent with responsibilities for providing infrastructure and implementing projects. It was officially launched in 1989 in Senegal as a private non-profit NGO for contract management in recognition of the inadequate capacity of the public sector in developing countries to deliver infrastructure programmes quickly and efficiently. It has a special legal status with exemption from government rules with respect to procurement, disbursement procedures, staffing and salary policies to encourage efficiency. AGETIP manages every aspect of a project based on a programme of small and medium size labour-based public works to take account of inadequate project identification capacity of local governments, the limited output of SMEs, constraints of local consulting engineers, architects, quantity surveyors, planners and other consultants, and abundant supply of labour (skilled, semi-skilled and unskilled).

The specific objectives of AGETIP are as follows:

(a) to explore the application of employment-intensive construction technologies in executing public works that are economically and socially beneficial and to create employment;
(b) to build local capacity by providing training to improve the operational efficiency of the local construction industry; and
(c) to improve the effectiveness of public institutions (local and central government) to assist them in building capacity and maintaining infrastructure assets.

There are usually several components involved. An investment component for construction, rehabilitation and maintenance of mainly physical and social infrastructure – roads, schools, health facilities, police and fire stations, courts, public parks, playgrounds and markets. This is made up of donor and counterpart funds from governments and municipalities to create a sense of project ownership. There are two types of operations: (1) public works including routine maintenance of roads, rehabilitation of drainage system and

urban infrastructure; and (2) public services, notably street cleaning, clearing up spoiled urban areas and waste collection and disposal. Another component is supervision and monitoring through
technical support (engineering design and construction project
management) and capacity building through training for the beneficiaries and small- and medium-sized enterprises. The main thrust of
this component is to

- improve public procurements and contract management
  practices;
- enhance the technical capacity of domestic construction
  contractors and consultants; and
- improve the management of infrastructure by local governments.

There is usually a third component – the study activities aimed at
investigating more specific issues such as improving procurement
practices, urban infrastructure maintenance management; the creation of mutual guarantee company and a lending instrument for
municipalities, private sector participation as a support of the project in general and the preparation of the urban environment project.
These studies are often aimed at building consensus by bringing
together the various stakeholders to discuss key issues.

The implementation of projects is a crucial stage. AGETIP hires
consultants to prepare technical engineering studies, designs and
bidding documents for projects and to supervise works. It issues an
invitation for bidding to contractors to implement the projects, evaluates bids and signs contracts with the most responsive bidder. As
construction progresses, the quality of the works are evaluated and
payments are made to consultants and contractors until the final
handover of works. By February 1996, it was reported that AGETIP
had executed more than 1250 subprojects in Senegal and created
18 000 person/years of employment. The number of engineering
consulting firms and construction contractors also rose significantly.
Accounts were paid in an average of 3 days and the procurement
process was significantly reduced with majority of projects completed in less than 2 months. Labour costs to the total costs of subprojects varied between 23% and 28% above the threshold of 20%
but below the expected 33%. External audits are regularly conducted
unlike government agencies where donors do not always have the
power. This includes monthly or bimonthly management audits,
financial audits every 6 months and technical audits every year.
It has become known in Africa for managing labour-intensive public
works with transparent, streamlined and monitoring procedures
significantly improving the management of projects and donor
finances.

Author's commentary

AGETIP is a model for capacity building involving the public and private sectors to develop institutions/local resources to facilitate the implementation of infrastructure subprojects. Subprojects are designed and evaluated to ensure that significant employment is created and there are other socio-economic, cultural and financial benefits particularly to local government and SMEs. The critical success factors are the result-focused management culture, often with strong political support and independence, streamlined and transparent procedures, highly motivated staff, financial autonomy and the training provided for its partners. AGETIP's institutional impact is substantial in building local and regional capacity. The approach has been successfully replicated in other African countries after the pioneering example in Senegal. Such agencies have forged networks through AFRICATIP and other social funds in Latin America. The AGETIP model is important for small-scale development of contractors and consultants as it creates a demand and market for the services of small contracting and consulting firms by freeing up procurement by means of dividing works into small contracts and providing quick payments. Initially set up as a temporary agency, it has now been given responsibilities for design and construction management in a large number of other donor funded projects, and other functions beyond contract management. AGETIPs procurement practices seem to be leading to changes in government procurement procedures. For example, a public procurement agency has recently been established in The Gambia to manage all other government procurement activities, apart from those civil works procurement currently managed by AGETIP (GAMWORKS).

However, there are problems with monopoly situation in construction management but only for social projects as other non-social projects, including large ones are subject to competitive tendering in the usual way. There are also concerns about beneficiaries' involvement and long-term viability. They should be more closely involved in project identification and planning and there is also a need to explore the role of AGETIP in the operation and maintenance of infrastructure facilities to ensure sustainability. It will also be necessary to collect information on economic and social impact (e.g. permanent employment) beyond temporary jobs created.

Case study 5.2: The Skye Bridge, Scotland

The Skye Bridge was one of the first PFI projects to be completed in the United Kingdom. The intention was to replace the short ferry crossing between Kyle of Lochalsh and Kyleakin on the Isle of Skye.

The box girder prestressed concrete bridge was constructed by the free cantilever method as a cantilever structure. The main span is 250 m and the total length of the bridge is 570 m. The design of the bridge was carefully developed to blend in with its picturesque Scottish setting and comprehensive studies were made on the surrounding environment.

A concessionaire was appointed by competition in 1991 and works were completed in 1995. The cost of the project was £39 million (at 1991 prices) and £15 million was provided from public funds. The remaining £24 million was the agreed cost to be recovered through tolling by the concessionaire. Once the bridge is fully paid off, the agreement provides for toll charges to be withdrawn.

The one-way toll charge was estimated to be 5.20 GBP which was 0.20 GBP cheaper than the previous ferry toll. However, the toll made the Skye Bridge the most expensive toll bridge in Europe. This created enormous local resistance, especially as the existing vehicle and passenger ferry service was to be withdrawn, immediately the bridge was open leaving locals and tourists no alternative but to pay the tolls if they wanted to travel between the Isle of Skye and the Scottish mainland. The eventual toll charge for a one-way trip was set at 4.70 GBP and locals could enjoy a concessionary rate set at 3.00 GBP, provided that ten tickets were purchased at a time.

Once the bridge was open, a campaign was mounted by the local population, some of whom refused to pay tolls and were duly prosecuted by the Sherriff's Office. This campaign has been very bitter and protracted with protesters complaining about a loss of jobs caused by tourists refusing to pay the toll and visit the Isle of Skye. Several studies have been undertaken and it is claimed that the cost of collecting the toll amounts to 3.00 GBP and together with the loss in local revenue more than justifies the cost of buying out the concession. Other studies claim that the tolls charged up to mid 2004 have already more than paid for the cost of the bridge as originally agreed and there can be no justification for the continuation of tolls. The Scottish Executive is now working towards ending the toll regime after years of pressure from campaigners who say it is costing Skye 5 million GBP per year and more than 200 jobs.

Author's commentary

The rationale for providing a fixed link between the Isle of Skye and Scottish mainland was without question the right strategy to follow. The provision of a safe all weather bridge capable of providing fast and convenient passage with the potential to regenerate substantial improvements to tourism and revenue are major benefits. However, it would seem that insufficient thought was given to the socio-cultural, political and economic impact i.e. livelihoods of local people, effects on tourism and toll fee. The amount of the toll was high bearing in mind that the bridge has the potential to substantially increase user volume when compared to the restrictions imposed by the ferry service. In retrospect it may have been better to have extended the concessionary period to substantially reduce toll charges taking into account the monopolistic situation created by the bridge. This case study provides an excellent example of the long-term implications of the concessionary agreement and the adverse affects if mistakes are made.

Summary

This chapter has demonstrated the need for two levels of evaluation to be conducted in a complementary way. First, a macro evaluation at the national level focusing on the overall goal of a project in terms of its broad policy impact or key contextual variables such as economic, socio-cultural, institutional, political and environmental indicators. Second, a micro evaluation concerned with strategic issues relating to the viability of projects within the context of specific objectives. The project-specific variables include design choice, construction methods, size and type of infrastructure development. The importance of analysing the resource implications of infrastructure projects, and the management of resources to improve the delivery of sustainable infrastructure assets and services are discussed. An outline of different types of resource management models is provided with the underlying principles, strengths and weaknesses of each type of model. It is argued that the InfORMED approach underpinned by systemic principles is a powerful tool for evaluating projects at both the contextual (macro) and project-specific (micro) levels. It facilitates the selection of infrastructure projects consistent with policy and strategic objectives with their full resource implications. The chapter concludes with two case studies illustrating the principles discussed in the chapter. The first case relates to the AGETIP concept illustrating the specific circumstances where project (micro) factors such as design, type of infrastructure and construction variables may have to be altered and monitored to increase or maximise the level of socio-cultural and economic impact i.e. local employment, income creation and facilitate institution building and resource development. The second case focusing on the Skye Bridge project emphasises the need to consider a wide range of socio-cultural, political and economic factors in advance to arrive at an optimum solution for all concerned. It is further argued that effective partnership between national and local governments and communities is the key to the effective delivery of infrastructure policies.

References

Al-Mufti, M.A. and Cochrane, S.R. (1987), '*Construction programmes in development planning*', In P.R. Lansley and P.A. Harlow (eds.),

Managing Construction Worldwide, Vol. 1, E & FN Spon, London, pp. 190–198.

Amjadi, A. and Yeats, A. (1995), 'Have transport costs contributed to the relative decline of sub-saharan African exports? Some preliminary empirical evidence', Policy Research Working Paper, 1559, World Bank, Washington, D.C.

Ball, M. and Wood, A. (1995), 'How many jobs does construction expenditure generate', Construction Management and Economics, 13, 307–318.

Carrillo, P. (1996), 'Technology transfer on joint venture projects in developing countries', Construction Management and Economics, 14, 45–54.

Drewer, S. (1997), 'Construction and development: further reflections on the work of Duccio Turin', First International Conference on Construction Industry Development: Building the Future Together, National University of Singapore, 9–11 December.

Egan, J. (1998), 'Rethinking construction: report of the construction task force on the scope for improving the quality and efficiency of the UK construction industry', Department of the Environment, Transport and the Regions, London.

Hillebrandt, P.M. and Meikle, J. (1985), 'Resource planning for construction', Construction Management and Economics, 3, 249–263.

Lemessany, J. and Clapp, M.A. (1978), 'Resource inputs to construction: the requirements of house building', Building Research Establishment, Current paper 76/78.

Miller, J.B. and Evje, R.H. (1999), 'The practical application of delivery methods to project portfolios', Construction Management and Economics, 17, 669–677.

McCutcheon, R.T. (2001), 'Employment generation in public works: recent South African experience', Construction Management and Economics, 19, 275–284.

Morton, R. and Jaggar, D. (1995), 'Design and economics of building', E & FN Spon, London.

Operations Evaluation Department (OED) (1997), 'A success and challenge: AGETIP in Senegal, OED Precis', No. 148, World Bank Washington, D.C.

Pean, L. and Watson, P. (1993), 'Promotion of small-scale enterprises in senegal's building and construction sector: the "AGETIP" experience', In 'New Directions in Donor Assistance to Microenterprises', Organisation for Economic Co-operation and Development, Paris.

Robinson, H.S. (2000), 'A critical systems approach to infrastructure investment and resource management in developing countries: the InFORMED approach', Unpublished PhD thesis, South Bank University, London.

Rogers, M. (2001), 'Engineering project appraisal', Blackwell, Oxford.

Rosenfeld, Y. and Warszawski, A. (1993), 'Forecasting methodology of national demand for construction labour', Construction Management and Economics, 11, 18–29.

Strassman, W.P. (1970), '*The construction sector in economic development*', Scottish Journal of Political Economy, 17(3), 391–409.

Turin, D.A. (1978), '*Construction and Development*', Habitat International, 3(1/2), 33–45.

Uwakweh, B.O. and Maloney, W.F. (1991), '*Conceptual model for manpower planning for construction in developing countries*', Construction Management and Economics, 9, 451–465.

Chapter 6

Infrastructure project procurement

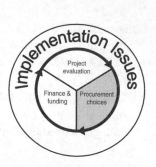

Introduction

The procurement of most public infrastructure facilities and services is traditionally the responsibility of government. This should be undertaken as a result of proper strategic planning to achieve national goals and aspirations aimed at the satisfaction of established needs. Moreover there should be in place a robust and objective prioritisation and allocation system designed to bring projects forward for consideration according to justified need.

The projects under scrutiny should have been rigorously assessed regarding their technical feasibility and financial viability in relation to the accrued benefits, both direct and indirect. Hence there should be primary alignment between strategic planning and the selection of infrastructure projects for implementation. The key issues regarding the evaluation of infrastructure projects have been discussed in the previous chapter. However, procurement is fundamental in the delivery or implementation of infrastructure projects. Procurement is the method of acquiring, securing or obtaining infrastructure assets, facilities or services. It involves the selection of various *actors* (public and private) and organisation of *activities* for infrastructure provision and production such as planning, financing, designing, construction, monitoring, regulation, operation and maintenance to facilitate the delivery of infrastructure services. Over the past 200 years governments have been faced with the dilemma of whether to procure projects by utilising public funds raised by taxation and borrowing, or by encouraging the investment of private funds for all or part of a project.

History indicates that the involvement of the private sector in the provision of infrastructure projects is by no means new. The introduction of turnpikes in the early 18th century was a primitive example of privatisation where landowners set up tolls for passage across their land by means of tracks and earth roads. In the United Kingdom the development of railways as an alternative means of transport to canals and roads during the 19th century led to complete privatisation of national railway provision involving local and regional companies. The subsequent lack of strategic planning, coupled with integration problems and standards of service led to local and regional railway networks being taken back into national public ownership during the 20th century. The latest development adopts a hybrid approach where the permanent rail infrastructure is publicly owned and the trains are operated by private companies.

Because of the vast expansion of infrastructure provision in the late 20th century and an increase in the public's expectation of service provision, the maintenance, repair and replacement of infrastructure has become progressively more expensive. Reluctance by the public to pay increased taxes to cover this expenditure has resulted in infrastructure not being adequately developed and maintained, thereby resulting in a drop in standards of service. Further, the provision of public funding for new infrastructure projects is now difficult to obtain. According to Levy (1996), the United States has consistently under invested in its infrastructure and he postulated that if the richest country in the World cannot afford to upgrade and maintain its infrastructure then what chance do less well off countries have of achieving this goal? This has led to the exploration of innovative procurement methods whereby infrastructure projects can be designed, constructed, operated and maintained with the support of the private sector. Despite previous failures of private ownership it is now perceived that the private sector has the potential to offer, in certain cases, more efficient management and resource utilisation compared to that of bureaucratic public bodies.

The provision of infrastructure has been much more difficult in developing countries where public and private funds are generally inadequate to fund large-scale infrastructure provision. As a rule infrastructure that can be funded indigenously is normally provided and controlled by the government. The usual sources for providing support funding for infrastructure projects in developing countries are the World Bank and regional development banks who provide loans repayable under very advantageous conditions. There may also be other contributions from developed countries that may be paid direct or through aid, bilateral or export promotion agencies.

This chapter identifies and evaluates various procurement methods. It starts with an overview of the key elements that influences procurement selection and examines the conditions under which infrastructure projects are procured. A framework is proposed for consideration of the most appropriate procurement options to be adopted, given the circumstances of individual projects. This concept is predicated on the basis that there will be situations where there is a place for procurement systems that are either funded publicly or privately, or by various combinations of both. Readers not familiar with the methods of procuring construction work should refer to Walker and Hampson (2003), Broome (2002), Rowlinson and McDermott (1998) and Akintoye et al. (2003).

The prime contribution of the chapter will be the identification and selection of the most appropriate procurement options to provide the best chance of project success. The special circumstances appertaining to developing

111

countries are taken into account and the processes for obtaining funds and assistance are described. Recommendations are included regarding those procurement methods that encourage and support the use and development of local indigenous human and appropriate non-human resources in the realisation of infrastructure projects.

Elements of an infrastructure procurement strategy

A strategic approach to the realisation of infrastructure projects has been identified and discussed in Chapter 4. The next step in the process is to develop a rationale for project implementation according to priorities established by integrated national, regional and local planning. Ideally, economies should be planned according to predetermined goals that provide a beacon to guide the appropriate selection of infrastructure projects to be brought forward for implementation. Such infrastructure projects may be broadly categorised as social, trade and technical (economic) and can be further broken down as defined in Chapter 2 and illustrated by Figure 2.2.

A combination of projects from these categories will provide infrastructure fundamental to economic and social development (Kessides, 1993). There should be an alignment between strategic planning and the procurement strategy to be implemented for the delivery of a portfolio of appropriate infrastructure projects. It may also be argued that there should be a mixture of publicly and privately funded infrastructure projects involving a range of delivery methods intended to implement strategy in accordance with budget and time-related targets. Connectivity between the development of national infrastructure and the development of the nation itself is essential. To achieve this it is necessary to have an adequate understanding of national culture and the manner in which communities interact together with an understanding of issues and conflicts. This in turn reinforces justification for investment in infrastructure and encourages greater efficiency and more sustainable solutions that provide better in-use-performance.

The essential elements of an effective infrastructure procurement strategy relate to what can be achieved within the bounds of technological feasibility, affordability and adequate standards of public safety. Within this context there needs to be commitment to fair competition, transparency and recognition of what is in the public interest. The key factors to consider are the nature of the cost and financial commitments involved, the division of responsibility between public and private sector, the actual infrastructure needs and characteristics of the projects, willingness to pay for infrastructure services, the need to encourage innovation in infrastructure development and the viability of the project.

Cost and financial considerations

Normally, infrastructure projects require substantial capital investment and once in operation they will consume resources to support their operation, repair and maintenance throughout the planned life cycle. Additional provision should be made for disposal, recycling and replacement. It is therefore essential to take a holistic view of these costs in order to make a comparison against benefits and the potential revenue from users. This approach will also establish whether projects can be sustained over their operational life cycle. Where projects are likely to be self-funding from user charges then they will be easier to justify and would be more likely to attract interest from private investors. There is also the procurement cost involved in putting a bid together for infrastructure projects. The cost associated with some procurement methods involving a wide range of activities (e.g. PFI) are more expensive than others (e.g. traditional, design and build) because of the higher transaction and logistic costs. However, this cost must be balanced against the benefits and other factors involved in the infrastructure development process.

Attention is required to the timescale of achievement related to project portfolios supported by sound financial analysis by the application of techniques and concepts e.g. discounted cash flow (DCF) and life cycle costing (LCC).

Division of responsibility

Infrastructure planning also requires a clear view of the interaction between public and private sector providers of infrastructure services. Responsibility for the provision of infrastructure falls in the first instance to national government, who may then devolve this responsibility to regional and local government, normally in accordance with policy and predetermined budget allocation. Government will make decisions according to what it perceives to be in the best interest of the nation and the well-being of its inhabitants. A central government agency or department usually controls schemes of national importance, whereas local government departments will deal with smaller infrastructure projects applicable to regions or districts. Government may therefore be seen as a client that has extensive responsibilities, some of which are summarised as follows:

- National strategic planning.
- National security and the maintenance of essential services.
- Development of ordnance or regulation that defines standards and their implementation.
- Representation of the best interest of the population at large.
- Action as an industry regulator to encourage industrial development and investment according to desired consumption levels. In this process efficiency and affordability should be linked to quality of service and safety.
- Inspection and enforcement.
- Facilitation and incentives utilising tax breaks, provision of land guarantees, etc.

However, there is a growing recognition that some nationalised industries wholly owned by the public sector would be more efficient under private control. The rationale behind this argument relates to a general lack of incentive for publicly controlled organisations to become more efficient and to develop innovative thinking to improve efficiency. Hence a worldwide trend has developed to sell off whole industries such as electricity, gas, water, telecommunications and railway infrastructure. Clearly, these industries are of strategic importance to the operation and well-being of any nation; therefore, government must retain the obligation to regulate and inspect standards of performance and service. In the event of default, government normally retains the ultimate sanction to take these companies back into public ownership and reserves the right to find another private investor.

Needs and project characteristics

Normally government will have a portfolio of projects that can be brought forward according to the need and availability of finance to support them from inception to full operation. Each project should be assessed regarding its sustainable operation throughout the planned operational life cycle and an evaluation should be made of the consumption of non-replaceable resources, pollution and the potential to recycle resources to reduce waste.

Project risk will require critical investigation and a comprehensive assessment should be made, which may impact on the selection of design and the technologies to be used, together with the method of project procurement. Where projects prove to be very risk prone then fundamental questions should be raised about viability and desirability. There is a need for the providers of infrastructure, at all levels, to interpret need in support of socio-economic, environmental, institutional and political aspirations. Related goals will be logically developed in the form of strategies and plans that conform to the overall direction provided by government. National and regional land use planning and the interpretation of demographic trends will play a key part in developing strategies that stand up to detailed scrutiny and robust questioning aimed at justifying the approach adopted. From these base objectives, budgets will be generated and responsibilities identified for implementation.

Over previous years it has been demonstrated that the public will expect a progressively higher level of service provided by infrastructure in line with advances and other developments in society.

Willingness to pay

There is a natural reluctance to pay charges for infrastructure services that were previously provided free and there is also strong evidence that where charges are considered to be excessive, then users will seek alternatives with the prospect of a fall in the volume of use. This may be evident when expressway charges are set too high. In the event that infrastructure services are provided free then all the funding required will need to be provided from government, normally through taxation and borrowing. In developed countries this has a tendency to create a dilemma because as infrastructure provision grows so does the cost of operation, maintenance and repair, which over the complete project life cycle could be up to ten times or more of the capital cost. The scale of this requirement very often exceeds the capacity of governments to fully fund upkeep and as a result there is a danger that infrastructure falls into a state of disrepair. Furthermore, the public are not generally in favour of rises in taxation to pay for more infrastructure. Hence, other ways need to be established where additional revenue can be generated from users. This requires a liberal approach to the adoption of infrastructure delivery policies that provide for the adoption of a flexible procurement framework by taking into account the differing circumstances and requirements of individual projects. At the extremes of this framework will on the one hand be infrastructure projects that are totally funded from the public purse and undertaken entirely by public works departments; and at the other will be projects that are wholly funded from private sources and are operated autonomously by privately owned companies. It is becoming increasingly apparent that there is a move away from total public ownership and there has been a significant trend towards the introduction of public and private participation in a variety of forms.

Innovation and new product development

The encouragement of innovation and technological advances is a key action that contributes to the performance and affordability of all types of infrastructure. Therefore, as a matter of necessity, bid documentation should contain provision whereby bidders are given the opportunity to introduce new ideas, concepts and technologies.

According to the New Oxford Dictionary *Innovation* is defined as:

'to make changes in something established, especially by introducing new methods and new ideas or products'.

The term innovation derives from the mid 16th century and is taken from *innovate* – 'renewed altered' and from the verb innovare, from in – 'into' + *novare* 'make new' (from *novus* 'new').

The cornerstones of innovation are *creativity* and *invention*. Further clarity can be found in the Collins English Dictionary where *creativity* is defined in three ways:

- possession of the ability or power to bring into existence;
- characterisation by originality of thought or inventiveness; and
- stimulation of the imagination beyond what already exists.

Similarly *invention* is defined as:

'the discovery or production of some new or improved process or mechanism that is useful but not obvious to experts skilled in a particular field'.

The nature of the process involved with design, construction, operation and management of infrastructure lends itself to incremental development and improvement, whereby over a period of time major step changes become apparent. However, there are instances, albeit less common, where 'breakthrough' occurs in technology or processes that alters completely previously held theories and paradigms.

New product and process development may also require specially commissioned research that may be defined as pure or applied. Pure research is unlikely to have any direct impact on the provision of infrastructure because it is concerned with the discovery and development of fundamental theories relating to the physical and natural environment and human nature. Applied research normally has a direct bearing on everyday life, including the development of new processes and technologies.

The definition of *research* is broad and may be stated as:

- a systematic investigation to establish facts and principles or to collect information on a subject;
- carrying out a critical investigation into a subject or problem; or
- a careful search or inquiry intended to discover new or old facts by the scientific study of a subject.

From the above definitions it can be deduced that innovation is not the same as research, although one may be derived from the other. It is possible to participate in innovation without research, especially in cases where discoveries are made by accident or chance, rather than the result of a planned programme of scientific study. Therefore, these terms need to be used in a precise manner to describe the true nature of innovation and research (I&R).

I&R plays an important part in the development of processes and technologies that contribute to the advancement of infrastructure provision. The pace of I&R relates primarily to the incentive to invest in new development; clients may bring this about, but equally it could be undertaken by commercial organisations to gain competitive advantage in the market for their products or services. Where a culture exists within the market for innovation and development there will be no alternative but for competitors to be constantly striving for new products and processes. Failure to react effectively will invariably lead to loss of market share and in the ultimate case business failure.

New developments are usually geared to increased levels of performance, greater efficiency, sustainability, conservation and the safety of occupants or users. Improvements may be introduced by new technologies, innovative design, re-engineering processes and the introduction of systems, standardisation and innovative project management. The development of more effective procurement processes can also greatly assist the efficient management and delivery of projects (Mody, 1996).

Because of the increasing popularity of public–private partnerships (PPP), greater attention is focused on life cycle performance together with associated costs. Consequently, procurement methods such as build operate and transfer (BOT) and private finance initiative (PFI) will be looking to accurately assess building performance and the risks involved, especially where materials, components and fabrications have been shown to be prone to failure. Where projects are expected to involve more risk due to such conditions as environmental exposure and heavy usage, or uncertain

116

socio-economic and political circumstances, then additional associated costs will need to be factored into the bidding process.

Project evaluation

Once a decision has been made to proceed to the initial stage of project, it will become necessary to consider how the project is to be funded. The key action will be to determine profitability and/or whether funds are to be sourced from the private or public sector, or a combination of both. Hence all possible sources should be investigated and evaluated according to the following factors:

- benefits and costs;
- potential for project success;
- degree of risk;
- procurement methods available;
- extent of incentives to encourage private investment; and
- relevant technical and design studies.

The outcome of this process will be a decision to proceed, or otherwise, with the design and preparation of specifications. Further feasibility studies will normally be carried out before the go ahead is given to select a suitable procurement strategy and selection procedure.

Before entering into the procurement process it is essential that the costs and benefits to be derived from infrastructure projects are properly considered. The value of each project will need to be assessed in relation to the holistic provision of services and the contribution provided, including indirect benefits to the wider environment, national economy and the quality of life. According to the nature of infrastructure projects they are usually long term and will therefore be reactive to macro economic planning and changes in government policy and society. Consequently arguments in favour of infrastructure development must be apparent, for example:

- significant contribution to economic development;
- potential for strong financial returns, thereby increasing the possibility of private investment; and
- high economic benefits, but low financial returns, which may mean the introduction of incentives to attract the right sort of investment.

It is worth bearing in mind that resources available for infrastructure development are limited and careful consideration must be given to an optimal portfolio of infrastructure projects that will produce the best overall result.

As a general rule where there is already high economic growth the demand for more and improved infrastructure will follow suit. Conversely, it may be desirable to induce economic growth by the introduction of more and better infrastructure. A comprehensive economic analysis will help to determine the best policy and course of action. Whatever the situation it

117

will be necessary to think well ahead using sound foresight and planning to develop the necessary long-term strategy to incrementally build up capacity. Impact on the environment should form part of this process whereby the perceived consequences to the bio-geophysical environment are understood to the extent that adverse affects concerning the quality of human life can be avoided or mitigated.

Drivers for infrastructure procurement selection

By definition, a client initiating an infrastructure project may be national or local government, or a private company. Either way the client will be looking to procure projects in the most efficient and appropriate manner to realise established goals. The manner in which they are to be procured will have a significant bearing on the success of the project and the extent to which life cycle goals are kept within budget. It is therefore important that the drivers for the selection of the procurement process are fully appreciated and understood. The selection of procurement option is influenced by key elements discussed in the previous sections and examples of the relationship between these key elements and selected drivers are summarised as follows:

Key elements	Examples of Drivers
Needs and project characteristics	– nature and size of the project – complexity – impact on national, reginal and local economies
Innovation/new product development	– necessity for new technologies, R & D – flexibility – speed
Division of responsibility	– impact on the environment and society – regulation of service, quality assurance – monopolistic service
Willingness to pay	– potential to generate revenue from users – income levels
Cost and financial considerations	– portfolio of serial projects – sustainability of services
Project viability	– profitability – economic returns – impact on environment and society

The relationships shown above are not exhaustive as such but illustrative of the different types of cause-effect relationships that influences procurement decisions. For example, complexity as a driver does not only impact on 'needs and project characteristics' but could stimulate the need for 'innovation and new product development'. Similarly, the 'potential to generate revenues from users' or income levels could affect the 'willingness to pay' for infrastructure services as well as the cost and financial considerations. The key elements of a procurement strategy and drivers which are intrinsically linked, directly or indirectly affects the degree of public control, the level of private sector involvement and funding, and the need for an integrated or fragmented approach in the delivery of infrastructure projects. For example, greater integration in design, construction and operational activities may be necessary to speed up implementation or to deal more efficiently with complexity in infrastructure projects. This integrated approach may also be required where there is a need for innovation and continuous product development as lessons learnt on the impact of design decisions during the operation of infrastructure facilities can be reused in the design of innovative infrastructure in the future. Similarly, a lower level of direct public control may be desired where there is drive for new product development to facilitate innovation in design, construction and operation of infrastructure facilities. Projects that have the potential to generate profit, with good economic returns and revenue from users willingness to pay for infrastructure services are likely to result in a higher level of private sector participation and funding. However, projects with certain characteristics relating to sensitive issues such as national security, public service obligations or associated with significant negative externalities are more likely to require an increased responsibility from the public sector and therefore some degree of direct control to mitigate adverse effects. Similarly, monopolistic infrastructure provision will also mean an increased responsibility for the public sector and some degree of direct public control in the procurement of such infrastructure services.

The relationship between these factors – degree of public control, level of private funding and level of integration are used in the development of a procurement framework discussed in detail in the next section.

Framework for procurement delivery

The delivery framework for the procurement of infrastructure projects should be based on a systemic approach that recognises the need for flexibility by allowing a wide range of procurement methods that can be adopted individually or in combination. Provision should be made to recognise varying degrees of public control and a range of solutions involving private participation. It is therefore reasonable to assume that if government requires more control then this will have a direct impact on the amount of public funds required. Given these circumstances the private sector will participate by bidding for contracts under the direct control of the public client. Where greater responsibility has been devolved to the private sector, then subject to an evaluation of acceptable risk, entrepreneurs will be more willing to consider the possibility of making the necessary investment (Miller, 2000). Account also needs to be taken of the amount of integration between the various actors in the process. Situations will vary from complete integration

119

at all levels to total fragmentation where different roles are conducted according to different paradigms that have little or no intended cross linkages. Supporters of a more fragmented approach will argue that there is higher potential for flexibility and hence there is likely to be greater capability to deal with expected dynamic events. Conversely, integration provides for seamless processes that eliminate duplication and disruptive inefficiencies, nevertheless meticulous planning is required to help mitigate uncertainty and reduce risk (Hudson et al., 1998). Projects characterised by a high degree of uncertainty and risk are more likely to be unsuited to a high degree of private participation. This is because private contractors will assess the risk with a view to adjusting their prices accordingly. In certain circumstances clients may find it more beneficial to take on the risk and then use their autonomy to minimise risk as the project progresses.

Using principles developed from Miller (2000) and Smith (1999) according to scales representing 'degree of integration' and the level of direct 'public control required' linked to level of private funding (see Figure 6.1).

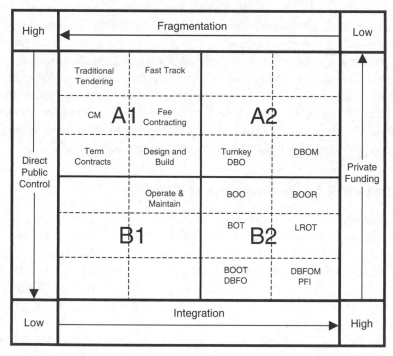

Figure 6.1   Infrastructure procurement framework (CM, Construction Management; DBO, Design Build and Operate; DBOM, Design Build Operate and Manage; BOO, Build Own and Operate; BOOR, Build Own Operate and Remove; BOT, Build Operate and Transfer; LROT, Lease, Renovate, Operate and Transfer; DBFO, Design Build Finance and Operate; DBFOM, Design Build Finance Operate and Manage; PFI, Private Finance Initiative; BOOT, Build Own Operate and Transfer).

Procurement methods

Framework sector A1

The methods described in this category rely on funding from government acting in the role of a client who is in direct control. The design and construction stages are separated from the operation and maintenance of the facility throughout its life cycle. A description of each procurement route identified is as follows:

(i) *Competitive tendering – The traditional method.* The responsibility for producing design and tender documentation rests with the client who will appoint a design team and in most cases a project manager who may also be the principal designer. Contractors will be invited to tender for the work by public advertisement, or from a short list of approved contractors. The contractor selected to undertake the project will usually be the one with the lowest price, although other factors may be taken into consideration such as time to completion and qualification to the tender price.

On completion of the project it will be formally handed over to the client for occupation. The liability for faulty workmanship will terminate at the end of the defects liability period, normally between 6 and 12 months after the agreed date of handover. The client will then take over responsibility for the operation and maintenance of the facility. A client controlled maintenance management team may achieve this or an independent facility management company may be engaged under a fixed term contract.

The traditional route is fragmented and has been subject to considerable criticism for its lack of incremental improvement and development. There has also been strong evidence of poor teamwork and damaging adversarial disputes. However, the traditional method does provide flexibility and it is usually easy to demonstrate public accountability through the selection of the lowest price.

(ii) *Design and build (D&B).* Some clients may not wish to become too involved in the design process but instead would prefer for delivery of the completed building to be in accordance with a brief normally prepared on behalf of the client. D&B therefore offers an integrated package that normally ensures excellent communication and co-ordination between the design and construction teams. The D&B contractor is usually appointed on the basis of prequalification after which an incremental process of negotiation leads to the agreement of a guaranteed maximum price for a finalised design and specification. This method has the advantage of mitigating the possibility of fragmentation and adversarial problems between the design team and the contractor. It is also argued that the project can be more efficiently delivered within a shorter time period by adopting more appropriate construction techniques. This method does not offer so much flexibility compared to that of the traditional approach and changes after the design has been agreed are usually more expensive to incorporate. Prime contracting is a further development of D&B where the integration of the whole D&B process is within a single supply chain. Hence the prime contractor may be defined as having overall responsibility for the delivery of the project, including the co-ordination and integration of consultants and contractors to efficiently fulfil the needs of the client, including the prediction of whole life cycle costs.

This category of procurement methods requires the client, on occupation, to make separate arrangements for the operation of the facility, together with maintenance and repair after the expiry of the defects guarantee period. D&B has been successfully applied to a generic range of construction projects; however, its application to certain large infrastructure projects is less advantageous because of the complexity and the significance of the cost involved with life cycle operations, maintenance, repair and replacement.

(iii) *Fast track (concurrent engineering)*. Where it is vitally important that a project is delivered in the shortest possible time, serious consideration should be given to the use of 'fast track' sometimes referred to as 'parallel working'. Design and construction activities are overlapped so that a facility is designed and constructed in stages or phases thereby compressing the project programme. This approach requires exceptional project management skills to ensure that disruptive elements do not cause serious non-productive time and consequential cost penalties that would destroy the purpose of the approach.

(iv) *Fee contracting*. This is an arrangement where the client engages a contractor on a fee basis for the management and delivery of the project. Essentially the client enters into a contract where there is agreement to reimburse the contractor for the full cost of the project, plus a percentage of the final contract sum to cover the fee for supervising the work. Fee contracting is a high-risk procurement route where the client carries the full responsibility for cost overruns. This route may be appropriate where a client has an expert in-house team or where there is sufficient uncertainty that will discourage contractors to submit competitive tenders for the work. The more successful variants of fee contracting have been developed around a strong partnership between the fee contractor and the client founded on understanding and trust generated over a series of projects.

The most common form of fee contracting is the management contracting procurement route where the client appoints a management contractor as a professional consultant at an early stage of the project. The management contractor then becomes part of the professional team and will support the design team with construction expertise aimed at improving efficiency. The provision of this service will be based on the expertise of the management team and their track record, which will be presented to the client prior to appointment. There may be a process of prequalification, prior to a competitive presentation before final appointment.

The management contractor will be expected to appoint works contractors who will undertake discrete packages of work. The contractual arrangement will be between the management contractor and the works contractor, however, it is important to realise that the client is duty bound to reimburse the management contractor for all costs incurred. Therefore, under this arrangement the client mainly carries the risk. This procurement route has largely fallen out of favour with clients in the global construction market, but where conditions are right it may provide an appropriate solution.

A major criticism of management contracting has been risk aversion by some clients who have attempted to unload risk to the management contractor. One way of doing this is by insisting that the reimbursement of package contractors will be subject to a maximum tender figure, hence any cost overrun, other than agreed variations, will be the responsibility of the management contractor. This has led to adversarial relationships, which have impacted on the popularity of this route in recent years.

(v) *Construction management.* This is a form of contracting where the client places direct contracts with each of the specialist contractors and suppliers. The expert management contractor is appointed as a consultant whose task will be to effectively manage and control the construction process, as well as integrating the design team. In this manner the client has more direct control over works contractors and this helps to mitigate the risk associated with management. This implies that the client is familiar with construction and has a close relationship with the professional team. The expert construction manager is therefore appointed as a consultant whose task will be to effectively manage and control the construction process, as well as integrating the design team. This procurement route shares many of the advantages and disadvantages that are claimed for management contracting. Overriding influences in the selection of this route concern the complexity of the project and the strength of the risk involved.

(vi) *Term contracts.* Term contracts are intended to create a basis for the reimbursement of maintenance and repair activities that are required for one or more premises occupied by the client. The contract is based on agreed system of rates and on-costs that are applied according to the need for work to be undertaken. This normally requires careful control by the client through a works officer who is responsible for authorising work and signing off work based on hours taken, or measurement against the schedule of rates contained in the contract.

With the exception of (vi), the outcome of the above procurement routes is the commissioned and completed facility. The client is then faced with the task of providing effective operation and control of the facility, together with maintenance and repair. This can be provided directly by an in-house team, or by a contractor who is an expert in the field of facilities management. Normally such contracts cover a predetermined period and will be subject to renegotiation or tender to secure a further term contract. As a consequence, the client is required to fund both the capital costs and subsequently the operational, repair and maintenance costs associated with the project.

A major factor will be the circumstances surrounding the requirement for the user(s) to pay rental or predetermined charges designed to create revenue. Under such conditions cash flow becomes very important and the revenue generated must be discounted to generate present values in order to provide a meaningful return on capital outlay.

Fragmentation does provide the client with greater procurement and decision-making flexibility, but it does impose greater demands on project management to ensure that integration is achieved through good co-ordination and communication between the participating parties. There is also the danger that whole life cycle considerations are not taken fully into account.

Framework sector A2

Retaining the principle of high public client control, greater integration may be sought by extending the procurement method to cover the operation and maintenance of the facility. Under this group of procurement methods initial planning and design criteria are supplied by the public client who will then appoint a single contractor to provide the design, build and operation of the facility (DBO). This arrangement can be extended to include maintenance (DBOM), thus providing a service with a high level of integration. The advantage with this approach is that the contractor is forced to consider and select the most beneficial solution. Financing can be from the public client or a series of cash payments, such as the right to recover charges from the users of the public facility. Off balance sheet finance is achieved by fees that are passed directly through the client (government) to the user, rather than involving an outlay of tax revenues. Where user revenues are insufficient then the difference will be made up by subsidies or shadow tolls. Hence the client retains revenue risk (Akintoye et al., 2003).

Turnkey is a variant of D&B that provides more integration, which as the term implies, delivers the completed building to the client for immediate occupation.

124

Framework sector B1

This sector represents higher fragmentation and a lower degree of public control. Only one procurement route is contained in this sector, namely that of pure operate and maintenance (O&M) where the client elects to outsource these tasks under a separate contract. While this type of contract integrates all post-occupation activities it is completely divorced from the design and construction processes. Furthermore, the client is not involved directly in day-to-day activities, although there is periodic control when contracts present themselves for renewal.

Framework sector B2

Where integration and private finance are the driving factors and the existence of less public control is not an issue then procurement methods need to be considered that have the ability to generate enough revenue to make private investment worthwhile. This is therefore the domain of 'PPP'. An important pre-requisite of this sector of procurement methods is the potential to gener-ate sufficient user revenues to provide the motivation for private investment against the prospect of acceptable returns given a level of perceived risk.

Concession-based models provide the necessary incentive for private investment by allowing the concessionaire to operate the facility in return for the right to collect user related revenues during the agreed concession-ary period. At the end of the concessionary period all operating rights and maintenance responsibilities revert to the government or public partner. It is optional as to whether legal ownership of the project rests with the con-cessionaire or not. In the majority of cases legal ownership terminates at the end of the construction period and the concessionaire retains only the right to operate the facility for the full concession period. It is possible to negotiate a further concessionary period if both parties agree. This process describes the standard 'BOT' process (Levy, 1996).

There are a number of models that have been developed over the past 25 years. The earliest models were known as 'build own operate and transfer' (BOOT) where ownership transfers to the concessionaire until the end of the concessionary period.

Due to the varying nature of infrastructure projects and in some cases the difficulty in predicting revenue stream it may be necessary to have some degree of flexibility during ownership transfers. There may also be cases where there should be no transfer of ownership. Because governments have had doubts about the desirability to cede the ownership of strategically important facilities, this method has become less popular and has been over-taken by other more recently developed variants. The practice of the client taking ownership at the end of the construction period allows the contrac-tor/developer to take responsibility for the maintenance and operation of the facility for an agreed concessionary period. Payment for the operation of the facility may comprise fixed or variable fees, or a combination of both. This method is normally referred to as 'BOT'.

Full privatisation requires a concession for a fixed period without the transfer of ownership to the public client. Hence, the concessionaire is responsible for design, funding, construction, operation and maintenance over the full concessionary period. On termination, the concessionary period may be extended or renegotiated. Alternatively, the facility may be sold, or in some cases decommissioned and removed, hence the terms:

- BOO (build, own and operate)
- BOOR (build, own, operate and remove).

Where the public client owns a facility requiring renovation or modernisation it is possible for the concessionaire to renegotiate a rental involving the obligation to upgrade the facility. In exchange for this investment a concessionary period is granted where income from the operation of the facility may be in the form of a service charge, together with user-related fees in certain cases. Normally at the end of the concessionary period ownership is transferred back to the public client. Hence, the title 'lease, renovate, operate and transfer' (LROT).

A more modern approach is to make the concessionaire responsible for:

- design and construction;
- operation, maintenance and repair; and
- provision of all the necessary finance.

Usually the public client provides the land/premises free of charge and cedes the right of operation to the concessionaire. In exchange for the provision of the infrastructure facility by the concessionaire, a basis will be agreed on how income is to be provided over the concessionary period. This may include income generated from users, fixed fees, performance-related shadow tolls or rent. No right of ownership of the land or the facility is given to the concessionaire over the concessionary period. Normally provision is made for the inspection of the facility prior to the termination of the concession to ensure that it is handed back in good condition. This approach is known as 'design, build, finance and operate' (DBFO) and it can be extended to include the management of the business conducted within the facility (DBFOM).

Other procurement considerations

Cash flow modelling is important to gain a better understanding of the relationship between costs and revenues, which will assist decision-making regarding capital budgets and expenditure patterns. The analysis should build a clear picture of the interaction between quality and cost covering the whole life cycle from project inception to decommissioning and disposal. Where appropriate, financial models may also be used to account for simultaneous project delivery.

In some cases it may be necessary to consider the possibility of adopting more than one delivery method for a single project. Hence, it is essential to compare and contrast project delivery methods, both singly and in combination, as deemed to be necessary to optimise the performance of the preferred procurement system.

It may be argued that revenues collected by government from taxation should not be used exclusively to support the provision, maintenance and repair of infrastructure. Therefore the strategy to be adopted should reflect the need to seek alternative sources of revenue from users as shown in Figure 6.2.

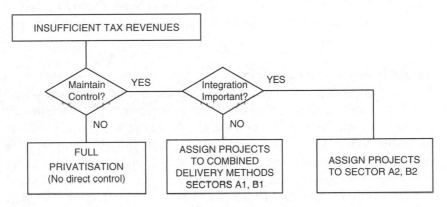

Figure 6.2   Strategic options.

Innovative procurement options

Concessionary agreements

A concessionary contract is enforceable in a court of law and will usually be between the host government/client and the concessionaire. The document is invariably extensive and will include appendices and will link to a web of subsidiary contracts.

The integrity and content of the agreement should cover the following major issues:

- description of the concession, its scope and duration;
- drawings, charts, schedules and specifications;
- the performance standards to be achieved as defined by benchmarks;
- guarantees and warranties against substandard performance;
- revenue agreements and other rights granted to the concessionaire;
- provision for changes in legislation and political decisions that may impact on the project;
- circumstances where the host government may take over the concession and the rights of the parties concerned;
- provision for termination of contract by either party; and
- provision for force majeure.

By their nature concessionary agreements must be detailed and are therefore invariably complex. However, too much detail restricts flexibility and places project management in a difficult position when innovative solutions and dynamic decision-making is required. One way of achieving this is to specify the standards of service required rather than that of the facility to be constructed.

Another difficulty may be experienced when the concessionaire's bankers require loans to be taken against certain security over the concessionary agreement in the event of default of the concessionaire. One way of dealing with this is to have a 'step in' agreement between the granter of the concession and the banker that allows for the bank to take over the concessionaire's responsibilities.

Private finance initiative (PFI)

PFI is a concessionary method developed in the United Kingdom, which transfers a significant proportion of infrastructure from capital expenditure to revenue expenditure. This process also transfers risk from the public to the private sector. As a consequence, public sector infrastructure projects can be brought forward even though public sector funding is not available. Companies wishing to bid for a PFI project are required to set up a special purpose vehicle (SPV). The SPV can be an existing construction company, a joint venture, a consortium or a specialist PFI company (Construction Industry Council, 1998). The preparation of the bid can be a lengthy and costly process and therefore governments require to give due consideration to the number of bids invited. A preferred bidder will be selected and a period of negotiation will take place to gain agreement on all aspects. Normally, it is expected that lenders would demand a cash (equity) contribution amounting to 5%–15% of the envisaged maximum cost of the project to be submitted by the bidding firm, the remainder being supplied by debt funding.

On appointment the preferred bidder becomes the concessionaire who will be responsible for designing and constructing the project according to agreed specifications and performance standards. The project will then be operated and managed by the concessionaire over the full concessionary period.

Critics of this procurement method point to complex partnering and structural arrangements, highly complex financial modelling and the dangers of administrative inefficiency. Furthermore, borrowing interest rates available to private organisations will be more than those available to the public sector. Against these misgivings is the potential of the private sector to innovate new solutions and its natural ability to generate value for money.

The critical issues for PFI rely on the public sector to only bring forward project proposals that have been well thought out and have viability and a genuine need. In addition governments must take a realistic view of risk assessment in that where risks are high then a reasonable price must be paid. There must also be public acceptance that PFI is value for money and projects must be effectively managed (PFI, 1996).

Partnering

The purpose of partnering is to remove inter-organisational barriers with a view to overcoming conflict and promoting project success. By increasing transparency in dealings and reducing impermeability, more flexible ways of working are created that overcome entrenched attitudes and encourage team-work. By building up trust between individuals and groups, players are more willing to institute initiatives and assume personal responsibility. These factors provide the foundation for effective partnering agreements between commercial organisations. Hellard (1995) summed up partnering in construction by highlighting the importance of an implied covenant of good faith based on a person's word. Partnering is therefore not a traditional contract, instead it is an agreement which sets out mutually agreed obligations that are not directly enforceable in a court of law. It is normal to write down the agreement in the form of an *aide memoire* that acts as a reminder for parties to the agreement. Sometimes this is done with the help of an external facilitator who will assist with the development and cohesion of the project team.

It is also necessary for both parties to understand each other's strengths and weaknesses and to have a clear understanding of respective duties and responsibilities. To provide a good chance of project success there must be a genuine desire for an all embracing winning solution based on agreed objectives and a competent assessment of risk throughout the project. There must also be a genuine and accurate appraisal of the economics of the project and likely returns on investment. The choice of the correct procurement strategy linked to strong project leadership will also be a key factor.

Partnerships involving infrastructure projects and concessionary agreements require a strong partnership between the client and the concessionaire. It should also be appreciated that the public client may consist of a number of different governments during the period of the concession, hence the need for strong public support from the outset. It is important that proper public accountability procedures are in place at all times. This is likely to constrain partnership at the bidding stage, however, once the concession or contract is agreed then the adoption of a true partnering approach from thereon is likely to be very beneficial.

Government bureaucracy and organisational structure does not rest well with partnering and it is therefore important to push for the implementation of integrated project teams that share the culture of the project and who are capable of building a shared perception of trust. Project goals should be shared and there should be a commitment to rapidly dealing with critical issues and events that require dynamic decision-making. Conflicting interests should be rooted out and any mutual lack of confidence by the partners in others should be dealt with in a timely manner. Expectations should not be exaggerated unnecessarily and political changes should be dealt with effectively. Finally, project management must be maintained at a high level of performance to mitigate the adverse effect of unexpected events and to correct adverse trends at an early stage.

Infrastructure procurement selection using a portfolio-based strategy

Due to the scale and longevity of infrastructure projects a number of vital issues need to be addressed. Invariably the most crucial of these relate to the competition for funds and the unpredictability of national and local government funding budgets, brought about individually or in combination by such circumstances as crisis, economic pressure and public safety. There should also be in place a system of long-term capital planning based on a competent assessment of the scale of existing infrastructure and its operational condition, maintenance and repair requirements and the likely period to replacement. This will provide a reliable evaluation of upcoming costs and budgetary requirements to maintain standards of service delivery. The above implies that there will always be more projects than the absolute capacity to meet all requirements. It is therefore essential for governments to have in place adequate criteria necessary to select those projects that will satisfy the most demanding needs (Smith, 1999).

There are certain key elements that comprise a robust procurement strategy and these will include:

- a comprehensive system to evaluate needs, both generally and for individual projects;
- potential providers and the scale of competition;
- the adoption of fair selection procedures and the degree of transparency used in their application;
- competent selection of the final design that incorporates innovation and the latest technologies;
- evaluation and implementation of performance standards that address life cycle requirements and sustainability issues;
- financial validation of the project over its whole life cycle;
- adequacy of investment to support the proposed service level and quality of the infrastructure to be provided; and
- the implementation of multiple delivery and financial methods to optimise the attainment of specific project needs.

The infrastructure portfolio strategy selected will have an impact and place constraints on the number of projects commissioned and the selection of procurement route. It is therefore necessary to categorise projects with regard to their importance and where they fit into the delivery programme. Less urgent projects will be scheduled for later implementation and will be ranked according to need. These projects will be subject to

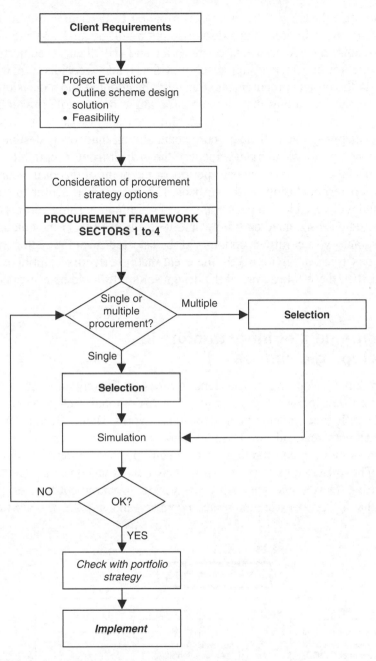

Figure 6.3   Project procurement selection process.

further prioritisation as other projects come on stream and in some cases where there is sufficient justification further delays may occur.

Projects budgeted to come on stream will be brought forward individually to be evaluated according to their previously determined feasibility. Client requirements will be brought together in the form of a brief that should contain all relevant and necessary information to determine one or more outline design solutions. Each design solution will be analysed and evaluated against the requirements of the brief and will then be subjected to assessment from which design selection will be made. Consideration will be given to the project procurement strategy to be adopted and a decision will be necessary regarding the adoption of a single or multiple procurement process. This activity covers all eventualities and allows the client to select the most appropriate holistic procurement solution that covers design, construction, operation and management of the project from inception to completion of the first concessionary period, or the point of disposal signalling the ultimate conclusion of the project. The preferred procurement route selected will be subject to further analysis by simulation using various models including discounted cash flow and life cycle costing. If the outcome is unsatisfactory then further options will be selected from the sector framework described in Figure 6.3. In the event that no effective solution can be found, then the whole nature of the design solution should be re-visited.

Procurement of infrastructure in developing countries

Given that developing countries tend to rely significantly on external funding from development agencies, there are additional factors to consider which influence procurement decisions. World Bank Group (WBG), together with regional development banks and aid agencies play a major part in supporting and funding infrastructure projects in developing countries. The WBG consists of five closely associated agencies as defined in Figure 6.4. Development agencies have various procurement rules or guidelines that affects the selection of actors/participants and the organisation of

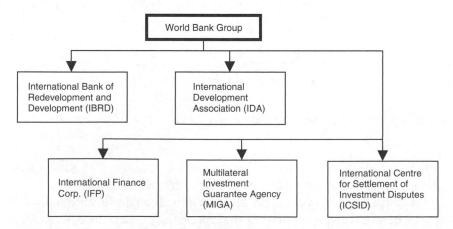

Figure 6.4   World Bank Group.

the activities and processes involved in infrastructure provision and production. These rules are applicable when applying for different types of funding relating to planning, design, construction and maintenance activities.

International procurement rules or guidelines

International procurement rules are aimed at selecting the most 'efficient' producer often relying on a large and diverse membership of countries to generate maximum competition. Figure 6.5 is a comparison of selected procurement rules of some development agencies. Some of these as well as other factors are briefly discussed below.

Selected procurement criteria	Eligibility to participate in procurement	Countries of operation or recipients	Mode of procurement	Bid Preference allowance (maximum)	Language of tender documents
World Bank	Member countries (regularly updated)	All member countries	ICB, other forms of competition	Local manufacturers, local contractors (7.5%) (only for countries below specified GNP)	English, French or Spanish
African Development Bank	Member countries	Developing member countries of Africa	ICB, other forms of competition	Local manufacturers (15%) local contractors (7.5%) regional co-operation among members (5% for works, 10% for manufacturing)	At least one of the two languages (English or French)
EBRD	All countries	Central and Eastern Europe, Commonwealth Independent States (CIS)	ICB, other forms of competition	No preference allowed	One of the Bank's working languages
Asian Development Bank	Member countries	Developing member countries of Asia	ICB, other forms of competition	Granted to local civil works contractors/manufacturers at the request of borrower but no specific level stated	English language
OPEC	All countries	Developing countries (except OPEC members)	ICB, other forms of competition	Local manufacturers (15%), in other developing countries (7.5%); local contractors (10%); contractors in other developing countries (5%)	Language used in international transaction
BADEA	All countries approved by the BADEA	African countries (except members of Arab league)	ICB, other forms of competition	Granted to Arab, African, joint venture Arab/African contractors where cost difference does not exceed 10%	
European Investment Bank	EU countries and beneficiary only	Mainly EU and ACP developing countries	ICB, other forms of competition	No preference allowed	One official EU language
Caribbean Development Bank	Member countries	Developing member countries of The Caribbean	ICB, other forms of competition	Manufacturers from Commonwealth Caribbean (15%); other regional members (7.5%); contractors from Commonwealth (7.5%)	English language

(a)

Figure 6.5(a) Comparison of selected procurement criteria of leading development agencies (OPEC, Organisation of Petroleum Exporting Countries, EBRD etc) to a separate page under 'List of Abbreviations' in the preliminary section of the book.

133

Selected procurement criteria	Advertisement and notification requirements	Period for submission of bids	Cofinancing and conditions for participation
World Bank	UN Development business, one newspaper of national circulation	At least 42 days, or 84 days for large/complex works	Participates in cofinancing but not in tied cofinancing funds; Bank procurement rules apply
African Development Bank	UN Development business, one newspaper of national circulation, local representatives of members, technical magazine for special works	At least 60 days, or 90 days for large/complex works	Procurement rules apply to own components in parallel financing; own procurement rules apply as condition in joint cofinancing
EBRD	UN Development business, one newspaper of national circulation, official gazette, official EC journal, procurement opportunities	At least 45 days, or 90 days for large/complex works	Participates in cofinancing but Bank's lending policy must be taken into account
Asian Development Bank	National newspapers of wide circulation (at least one English newspaper), local representatives of member countries, technical magazines for special works	At least 60 days for supply contracts and 90 days for civil works	Procurement rules apply to own components in parallel financing; own procurement rules apply as condition of joint cofinancing
OPEC	UN Development business, one newspaper of national circulation, countries local representatives, technical magazines for special works	At least 45 days, or 90 days for large/complex works	
BADEA	One local and international newspaper of wide circulation, at least one Arab newspaper, all Arab and African embassies	At least 6 weeks, or 12 weeks for large/complex works	Participates in cofinancing
European Investment Bank	Journal of European Communities, local newspapers	At least 6 weeks, or 12 weeks for large/complex works	Participates in cofinancing
Caribbean Development Bank	UN Development business, national newspaper of widest circulation, another newspaper outside recipient country, local representatives of member countries	At least 60 days or 90 days for complex works where advertised; 45 days where prequalification is used and 30 days where local bidding is used	Participates in cofinancing

(b)

Figure 6.5(b)   Comparison of selected procurement criteria of leading development agencies.

Eligibility considerations determine who can participate and, therefore affect the nature of competition in the procurement process. Some development agencies such as the Organisation of Petroleum Exporting Countries (OPEC), European Bank for Reconstruction and Development (EBRD) do not have any restriction to participation in the procurement process. Others such as the Arab Bank for Economic Development in Africa (BADEA), European Investment Bank (EIB), African Development Bank (AfDB), Asian Development Bank (AsDB) and Caribbean Development Bank (CDB) have restrictions.

The mode of procurement varies. International competitive bidding (ICB) is used by most funding agencies as the most transparent, efficient and economic way of procuring goods and services. Other forms of procurement such as limited international bidding (LIB) are used where the character of the contract is such that there are likely to be only limited suppliers.

National competitive bidding (NCB) or local competitive bidding as it is often called is used also where the character or size of the goods or works involved is unlikely to attract bids from outside the project holder's country. But the criteria for defining which project fits into the category of special circumstances vary from one funding agency to another.

Although a key philosophy in procurement is to select the most efficient producer, some development agencies have an overriding objective of encouraging the growth of indigenous capability in developing countries, by introducing a limited margin of preference in the evaluation of bids and by strengthening co-operation through joint-venture.

Joint-venture to facilitate technology transfer is encouraged by some funding agencies to enhance the expertise of personnel in developing countries. World Bank and CDB have provision for joint-venture in their procurement guidelines but have explicitly stated that any form of mandatory association of local and foreign firms is unacceptable. Joint-venture is also implied or encouraged by other development agencies such as BADEA, AfDB and AsDB.

The responsibility for the selection of consultants is normally that of the borrower or project holder. However, the procedures for arriving at a short-list is often determined, or influenced by funding agencies. World Bank require a short-list of three to six firms with wide geographical spread including at least one qualified firm from a developing country. The AfDB also require a short-list of wide geographical distribution of five to seven firms with at least one from a regional member country and no more than two firms of the same nationality.

The evaluation of the submitted bids of consultants is determined by a set of criteria and evaluation methodology but the criteria vary from one financier to another, and so are the methods of evaluation. Some financiers use the two-envelope system whereby bids are submitted in two separate envelopes containing the technical proposals and the financial (fee) proposals. There are three variants of the two-envelope system; the first method is the cost-weighted (quality- and cost-based) technique where the financial offers of the top ranked group of bidders (several firms) are opened and the price added on as an additional criteria to give a combined quality and cost evaluation; the second method is the non-cost weighted (quality-based) technique where the firm with the highest score for technical quality is invited for financial negotiations. The third variant is the threshold-based technique where a cut-off score for technical quality is established, and the firm with the least cost among those with the qualifying mark is selected. There are other methods such as evaluation based on 'qualifications and references' only, but are normally used for small assignments.

The choice of consultants and contractors for infrastructure design and construction are therefore influenced by key procurement rules. Procurement decisions, therefore, have implications for infrastructure development as they influence the likely choice of technology, materials and skills utilisation, local capacity building, recurrent cost for maintenance, operational efficiency and overall project sustainability. It is therefore important to know what is permitted under the various international procurement rules. Funding agencies usually attach conditions to any loans or grants and recipient governments are bound by their procurement rules. As a general rule the greater in involvement of the donor agency the greater the extent to which their rules will apply.

Donor agencies can be classified as multi-lateral e.g. WBG, bilateral e.g. Canadian International Development Agency (CIDA), Department for International Development UK (DFID), United States Agency for International Development (USAID), trust funds e.g. Public–Private Infrastructure Advisory Facility (PPIAF). Each will have some differences in procurement rules on a wide variety of issues that can be summarised under the following headings:

  (i) eligibility for funding;
 (ii) terms under which funds are provided;
(iii) requirements for competitive selection;
(iv) requirements for advertising and short listing;
 (v) evaluation of proposals; and
(vi) involvement of individuals.

Private participation in infrastructure

More recently agencies have recognised the importance of involving the private sector to encourage investment in infrastructure. The move away from purely public sector lending programmes has involved the donor agencies in playing a more pivotal role in assisting developing countries to redefine the role of the state and the restructuring of the market and institutions so that they can effectively involve the private sector. Hence, the introduction of Private Participation in Infrastructure (PPI), which requires the private sector to additionally manage, maintain, expand, operate and/or invest in infrastructure services e.g. energy, telecommunications, transport, water and sanitation (World Bank, 2001).

The prime reasons for the growing popularity of PPI are

* the increasing recognition by national, state and local governments that the private sector, under favourable conditions, can take responsibility for the risks involved with infrastructure projects;
* the increased pressure on public funding from other sources and the decrease in funds available for infrastructure development; and
* risks are spread more widely to include government, consumers, project sponsors, concessionaires, financiers and facility managers.

Involvement of the private sector will be relatively easy where there is sufficient opportunity to earn good returns on investment from revenue income provided by users. However, when revenue income is small or non-existent then other forms of income for private investors will need to be established. These may take the form of subsidies or shadow tolls intended to promote a reasonable chance of private sector profit. Herein is the fundamental problem for the poorest economics that are unable to sustain infrastructure beyond their financial limitations. It is therefore essential that PPI should be based on affordability and sustainability and it follows that adequate funds should be available to cover both capital and life cycle operating costs. Failure to achieve this will result in default to the old public sector donor model where inherent problems associated with large debts and the inability to maintain infrastructure results in premature decay and reduce project life span.

The four stages associated with the introduction of PPI include

   (i) establishing policy;
  (ii) setting the legal and regulatory framework;
 (iii) bidding; and
 (iv) project management.

As a general rule the lower the return the lower the risk; this tends to favour public sector involvement, whereas the private sector will engage when returns are higher and that the level of risk is acceptable. When selecting a method of procurement for PPI there will be a range of options to be considered, e.g.

- service contracts (fixed fee) 2–5 years;
- management contracts 3–5 years;
- leases 10–12 years;
- concessions 15–30 years;
- greenfield contracts 10 years – ever; and
- asset transfer, trade sale or flotation.

Considerable influence will occur over the procurement method according the degree of ownership and responsibility that government wishes to retain. Figure 6.6 provides the basis for selection.

It is important to appreciate that the process of procurement under agency arrangements will involve a framework of actions and procedures that must be followed to ensure consistency, fairness, transparency and efficiency in eliciting the best possible procurement outcome.

Selection of bidders

A key aspect of any procurement method is the selection of an appropriate bidder for infrastructure project implementation. According to the procurement method adopted, the next step is to decide upon the procedure to

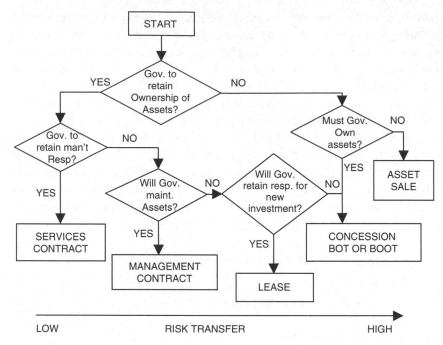

Figure 6.6   Choosing the appropriate method of PPI.

be adopted for the selection of bidders. The importance and scale of infrastructure projects makes it essential that the bid package is of good quality as well as logical in the development of its content. Experienced bidders will judge the quality of the bid package and the evaluation criteria as an indication of the feasibility of the project and government's intention to make it a success.

Options are available to clients ranging from public advertisement to direct negotiation with a single preferred bidder. There may also be government public accountability requirements and trading block rules, e.g. European Union requires work to be advertised publicly for tender or to elicit expressions of interest leading to various prequalification processes.

To reduce the amount of abortive effort involved with unsuccessful bids it is good practice to implement staged processes, whereby, the weakest applicants are eliminated early in the selection procedure with the minimum of wasted effort. The objective is to whittle down the serious contenders to a short-list who have the necessary expertise and resources to undertake the work. An alternative approach is to select the short-list from a list of approved contractors. Where the scale of project involves a joint venture or a consortium then the selection procedure should expedite the reduction of the short-list e.g. two to three bidders, thus providing a reasonable chance of success. It may be desirable for the client to make a contribution to the costs of the unsuccessful bidders and to pay compensation for the adoption of any of their ideas or innovations.

Information provided for the bidders

The type and extent of the information provided will depend on the nature and scope of the project, the extent to which previous investigations have been undertaken and the method of procurement selected. The aim should be to provide as much information as possible to reduce the number of unknown factors and hence the degree of risk involved; this will have a direct bearing on the bid price. There may also be provision for innovation and sufficient flexibility to encourage bidders to introduce new technologies and ideas to gain competitive advantage.

Typically the following information will be provided:

  (i) project description;
 (ii) feasibility studies, including preferred scheme design and outline drawings;
(iii) project design compendium incorporating the functional specification;
 (iv) environmental and geographical analysis;
  (v) geological data;
 (vi) existing services and utilities with growth forecasts etc.;
(vii) existing and projected volumes e.g. traffic flows, usage rates;
(viii) land use planning and transport strategies;
 (ix) project schedules; and
  (x) other relevant project information.

It is considered to be good practice to encourage bidders to seek further information and answers to unforeseen queries as they arise. Incentives should be offered to unsuccessful bidders in that if any of their innovative ideas are taken up, then an agreement should be reached concerning a fair level of reimbursement.

Evaluation of bids

The criteria upon which the bids will be evaluated must be properly developed to determine the winner. Normally, these will be published as part of the bid package to provide bidders with guidance regarding the preparation of their bids. The process of selection should be transparent and fair with feedback being given to unsuccessful candidates.

Usually the key project objective is broken down into the major issues to be addressed for its full and successful achievement. These might be as follows in the case of a project involving a concessionary agreement:

1. Financial viability
    – value for money
    – collection of revenue
    – robust nature of financial proposals
    – financial strength of the bidder

2. Excellence in design, technology and engineering
   - quality
   - cohesion and integration of the team
   - performance specifications
   - speed of construction
   - safety
3. Reliability
   - previous knowledge and experience
   - life cycle costs
   - security
   - arrangements for maintenance and upkeep
   - operational performance benchmarks.

The criteria will be subject to analysis by various techniques such as weighting and scoring systems and other methods as described in Chapter 7.

Case studies

Case study 6.1: Lane Cove Tunnel, Sydney, Australia

The Lane Cove Tunnel will link two of Sydney's major motorways, namely the Gore Hill freeway and the M2. It is located in a densely populated area and therefore there is a need to minimise adverse environmental impact. The tunnel is intended to provide easier access to the city centre from the north-western suburbs. The project consists of two 3.6 km tunnels, transit lanes, ramps and two new lanes across the Lane Cove River Bridge. There are also improved facilities for pedestrians and cyclists. At peak, it is expected that 9000 vehicles per hour will use the $(Aust) 1.1 billion tunnel which is due for completion in 2007.

The Client is the New South Wales Roads and Traffic Authority and the Lane Cove Tunnel Company (LCTC) has been engaged to design, construct, maintain and operate the tunnel for 33 years, after which it reverts to the ownership of the New South Wales State Government.

LCTC have appointed a joint venture between Thiess and John Holland to design and construct the project as part of a BOOT project. The Thiess John Holland JV was nominated as the preferred bidder in October 2003 and after meeting all conditions, the selection was confirmed and two months later both companies have combined their knowledge, skills and experience to comply with 259 conditions set by the NSW Ministry for Planning. These conditions include strict environmental monitoring and control during construction and detailed plans for managing water and air quality, dust and noise have been required prior to commencing on site.

The detailed urban and landscape draft framework plan has received sufficient approvals for tunnelling to commence in the Spring of 2004. A Construction Liaison Group (CLG) has been established to engage with local residents and businesses and to inform them of project progress. Steps have also been taken to enable queries and complaints to be dealt with effectively and efficiently. During the construction work CLG meetings are planned and a website is to be set up supported by a hotline and a bi-monthly newsletter.

Author's commentary
The Lane Cove Tunnel is a complex project undertaken in a densely populated area of Sydney. The selection of a design and build joint venture as part of a BOOT project by the Lane Cove Tunnel Company demonstrates the care needed to bring together the right knowledge, skills and experience to successfully complete the design and construction of the project in an environmentally friendly manner. A key factor in the appoint of the Thiess John Holland JV was their previous experience of large local infrastructure projects, including the Sydney Harbour Tunnel, North Side Storage Tunnel, the Melbourne City Link and the Chatswood to Epping rail line.

Case study 6.2: M77 Fenwick to Malletshaugh/ Glasgow Orbital, Scotland

The project valued at £132 million includes 15.2 km of two-lane motorway on the M77 and 9.4 km of dual carriageway for the Glasgow South Orbital (GSO), together with 800 m on the A726 junction with the GSO.

Balfour Beatty is the co-promoter and 67% investor in Connect the concessionaire who has been awarded a 32 year DBFO contract by the Scottish Executive and the East Renfrewshire Council. The DBFO Contract is the principal project agreement and it set out the obligations of both parties in relation to the project.

Under the construction contract between Connect and Balfour Beatty Civil Engineering (BBCEL), the latter assumes all design and construction obligations contained in the DBFO contract and is responsible for the management and obligations of construction subcontractors. The construction programme requires that the main M77 and the GSO elements are carried out over a period of 2 years and a further 4 months is allocated to complete the detrunking works to the existing A77 and associated roads.

Connect is responsible for all routine operational and maintenance obligations, including periodic major life cycle maintenance of the project roads and the replacement of worn-out elements. Activities and responsibilities are co-ordinated and managed by the Operations and Maintenance Unit of Connect.

Author's commentary

A crucial issue for this type of project is the generation of an adequate income stream over the 32-year period of the DBFO contract. Therefore financial agreement by parties requires closure before finalising the contract. Where revenue is raised by means of user tolls then great dependence and hence risk is placed on estimating correct future traffic volumes. However, where highways are toll free then the client will be required to provide periodic revenue payments to the concessionaire. These may be based on usage and or a system of shadow tolls.

Summary

Fundamental to the selection of an appropriate procurement method is the degree to which the private sector is to be involved in the financing, design, construction, maintenance and operation of an infrastructure project. Closely associated with this is the extent to which integration is necessary and the amount of flexibility required to achieve a coherent process throughout the life cycle of a project. Infrastructure projects that require flexibility in the system of delivery and have only limited capacity to generate low user revenues will be more likely to be suited to traditional methods of procurement that involve the private sector in competitive tendering, design and build and fee contracting. Under these circumstances the decision to retain ownership, or to lease to an operator need not be taken until after the design has been finalised and construction work has commenced. Such procurement methods imply that the majority of funding will be provided by the public sector in the first instance.

The sectoral framework proposed in this chapter describes how procurement methods can be progressively oriented to lever in private funding in exchange for concessions that deliver user revenues, rent and subsidies. The ultimate case of privatisation is where public assets or facilities are sold off to the highest bidder(s), thus their operation, subject to prior conditions, is at the discretion of the private owner. It is however very unlikely that governments will be prepared to relinquish the control of strategically important infrastructure without ultimate sanction or regulation over prices and quality of service. In this manner, expenditure is off balance sheet and therefore reduces the public sector borrowing requirement.

In the event of serious failure by a concessionaire, the client will invariably insist on the right to remove a concessionaire and to find a suitable replacement. The main advantage of granting concessions is that projects can be brought forward without the necessary public funds being available and risk is transferred to the private sector.

The adoption of a procurement strategy that fully considers the delivery of a portfolio of prioritised projects is of crucial importance. Only projects that can demonstrate real need and strong public support should be considered for implementation. Ideally such projects should have a strong revenue stream, but this will not always be the case. Where revenue income is small or non-existent justification for the expenditure of public money will need to be demonstrated to show that commensurate benefit is brought to the economy together with improvement in the living standards and well-being experienced by inhabitants within society. Therefore, robust and careful consideration must be given to indirect benefits that are to be derived. Security and public safety considerations will always be a major factor and risk assessment will feature strongly in feasibility studies that play a major part in the validation and realisation of projects.

Innovation, research and new product and process development play a key role in quality and performance improvement and will be major considerations in the development of competitive advantage amongst bidders. Bid packages should contain provision for the development of new ideas and technologies that contribute to the process of continuous improvement. In particular, issues revolving around sustainability and conservation should be given due attention to reduce the usage rate of the Earth's non-replaceable resources.

The selection of bidders should be seen as fair and transparent, especially in the evaluation of bids against stated criteria. The number of bidders should be carefully controlled to reduce unnecessary expenditure for those who are unsuccessful. A staged process of prequalification represents good practice by eliminating the weakest bids early in the process, thus protecting public interest against those bidders who are unlikely to produce the best solution. The final short-list of bidders should be as small as possible and in instances where large projects are concerned the majority of the final negotiations should be with the preferred bidder who will carry the heaviest bidding costs. In some instances it may be advantageous in the long term to mitigate the costs of unsuccessful bidders who may be required to enthusiastically bid for future projects.

The next chapter is concerned with the financing of infrastructure projects and the assessment of the risk involved.

References

Akintoye, A., Beck, M. and Hardcastle, C. (2003), '*Public private partnerships*', Blackwell Science.

Broome, J. (2002), '*Procurement routes for partnering: a practical guide*', Thomas Telford.

Construction Industry Council (1998), '*Constructors' guide to PFI*', Thomas Telford, London.

Hellard, R.B. (1995), '*Project partnering and practice*', Thomas Telford, London.

Hudson, W., Haas, R. and Uddin, W. (1998), '*Infrastructure management integrating design, construction, maintenance, rehabilitation and renovation*', McGraw-Hill, New York.

Kessides, C. (1993), '*The contributions of infrastructure to economic development: a review of experience and policy implications*', World Bank discussion paper No 213, World Bank, Washington, D.C.

Levy, S.M. (1996), '*BOT: paving the way for tomorrow's infrastructure*', Wiley.

Miller, J.B. (2000), '*Principles and practice of public and private infrastructure delivery*', Kluwer Academic Publications, Dordrecht.

Mody, A. (1996), '*Infrastructure delivery: new ideas, big gains, no panaceas. Infrastructure: private initiative and the public good*'. A. Mody (ed.), World Bank, Washington, D.C.

'*PFI BOT Promotion*', (1996), Project Finance International, No 93, March.

Rowlinson, S. and McDermott, P. (1998), '*Procurement systems: a guide to best practice in construction*', E & FN Spon, London.

Smith, A.J. (1999), '*Privatised infrastructure: the role of government*', Thomas Telford, London.

Walker, D. and Hampson, K. (2003), '*Procurement strategies: relationship-based approach*', Blackwell Science.

World Bank (2001), '*Toolkit: a guide for hiring and managing advisors for private participation in infrastructure*', World Bank Publication (ISBN 0-8213-4985-6).

Chapter 7

The financing of infrastructure projects

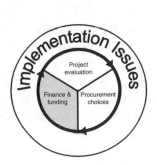

Introduction

The demand for infrastructure development and the maintenance of existing infrastructure caused by economic growth and population increase has, in many instances, overtaken the capacity of national governments to provide the necessary finance. As a consequence, it is possible that economic constraint may be caused by the provision of inadequate infrastructure. To greater or lesser extent, this applies to virtually every country in the World, including the United States of America. It is therefore logical that infrastructure previously provided free of charge is now coming under increasing pressure to generate additional revenue through the imposition of user charges in the form of tolls, fees and tariffs.

There has also been a widespread sell-off of public utilities to the private sector such as power generation, water supply, sewage treatment, refuse disposal and telecommunications. The flotation of public corporations has effectively transferred these organisations to private ownership by subscription to shares offered for sale on the stock market. Because these utilities are fundamental to public well-being and safety, government still has responsibility for their effective and efficient operation. Hence, these newly privatised organisations are subject to regulation and are required to justify to government charges and any price increases proposed alongside a range of ethical, safety and performance standards. The alternative is for government to establish a wholly publicly owned corporation that is responsible for charging users and then using the proceeds to operate and maintain the facility, as in the case of the French auto routes (Farrall, 1999). A solution may be sought somewhere in between wholly private and public ownership, where government has a role to play.

It has also been realised that government responsibility to provide transport infrastructure such as roads, railways, airports, health care, education and some forms of accommodation, i.e. prisons, university halls of residence can be outsourced to the private sector, whereby the revenue generated over a substantial period of time can be used to finance immediate capital expenditure and the full costs associated with the operation of an infrastructure facility over a concessionary period (Miller, 2000). This concept has considerable advantages for government in that public expenditure can be reduced and the need to raise funds by means of taxation and public borrowing is considerably mitigated. Consequently, government can take credit in the provision of new roads, hospitals and schools without having to find immediate funding from public sources. Critics of this approach argue that it is more expensive for the private sector to borrow money compared with government, however this is countered by the claim that private organisations are inherently more productive and their focused policies and management strategies are geared to the introduction of innovation and new technologies to achieve greater efficiency. Despite the advantages of 'public–private partnership' (PPP) not all projects are suited to this type of funding, since the income revenues generated may be inadequate, or the risks are so great that the cost of private funding far outweighs the benefits.

This chapter examines the options available for the financing of infrastructure projects and the techniques for economic and financial evaluation. The various methods available for funding projects are described and explained within the context of public and private project proposals. Projects are classified as being either directly or indirectly funded and are related to the procurement routes covered in Chapter 6. A selection of project financial appraisal techniques is examined to assess the feasibility of infrastructure projects. Specific attention is given to life cycle performance, capital and operating costs, together with, the ability to generate revenue income and the appraisal of uncertainty and risks. Methods of risk assessment are described and evaluated in the medium and long term taking into account political, economic and commercial factors, including new product development and innovation.

The bid evaluation process is described and the methods used for the financial appraisal of bidders are explained. The chapter concludes with the need for financial monitoring and control of infrastructure projects, both during the construction stage and during post occupation.

Financing policy and strategy

Infrastructure provision should be viewed as a means to facilitate economic growth or as a means to accommodate needs generated by economic development. The latter case is easier to justify because wealth will already be in the process of generation and financial resources should be easier to obtain. In reviewing infrastructure financing options, consideration should be given to sustainability and conservation issues discussed in depth in Chapter 9. Other environmental, social and political factors will also need to be taken into account. Reference should be made to Chapters 3 and 4 where the context for the selection of policies is described and the strategy framework for infrastructure implementation is explained. An overriding necessity will be that infrastructure must be feasible and affordable within resource and other constraints. Life cycle financial imperatives and related sustainability needs will play a vital part in the consideration and determination of financial strategies and these should be assessed in relation to their potential to generate user income.

Democratic national governments have ultimate responsibility for the implementation of policies to fulfil the pledges made to their respective electorates and to take action concerning public well-being and safety as the need arises. It is therefore essential that policies are converted into strategic and financial plans/budgets that identify national and regional economic, social and cultural development needs. National government normally devolves responsibilities to local government and to national agencies and these are supported by the distribution of public funds in the form of annual budgets that are normally ring fenced and subject to specific rules regarding the manner of their expenditure.

The allocation and distribution of public funds is therefore vitally important and will be justified according to relative need and priority. Government has a key role to play in this process by determining the manner in which public funds are allocated in support of its policies. The national budget will normally be identified under a number of headings that become the responsibility of government ministers and their respective departments. Typically departments concerned with the environment, transport, health, education and power generation will have annual budgets that include sums for infrastructure projects. The size of these budgets will depend on many factors, however the underlying influence will be economic wealth and the relative importance to future national wealth and prosperity.

Government therefore has a key role in the identification and prioritisation of infrastructure projects and subsequently the facilitation of their delivery and operation. It is now normally the case that public funds are insufficient to finance all the demands for new or replacement infrastructure. Further, as global economic development has continued to increase over the last century the amount of infrastructure requiring maintenance and repair has grown in a similar fashion. This coupled with the demand for new infrastructure to support economic growth has created an increasing gap between demand and the supply of public funds for this purpose.

The adoption of a dual-track strategy where private as well as public funds are sought by governments to support the provision of infrastructure projects is not new. As long ago as the early 1800s, the United States government was highly dependent on private investors for the development of new infrastructure and this practice has prevailed to the present day. It is therefore logical to deduce that governments today face similar problems as in the past, the main difference being the growing body of knowledge and experience available. This is especially the case in relation to technological

147

advancement and innovation in project delivery methods and systems of financing projects. As previously mentioned, infrastructure projects should be categorised according to whether their nature and size is suitable for public or private finance, or a combination of both. This approach may help to establish infrastructure portfolios that have similar characteristics, thus facilitating the development of performance benchmarks against which efficiency can be determined.

There should be alignment between strategic economic planning and the implementation strategy for infrastructure projects. There should also be secondary alignment with suitable procurement methods and the provision of finance. In this manner an appropriate balance can be struck between private and public funding, provided that selected projects display the following prerequisites:

- connectivity between the development of infrastructure and national requirements;
- instigators of the project understand the environmental and social context;
- the rationale for individual projects has broad support;
- support is available from investors who are prepared to risk their own capital and act from a position of financial strength; and
- returns are perceived to be good, given the application of discounted cash flow (DCF) analysis and risk appraisal.

The essential elements that should be contained within the infrastructure strategy depend on government having a clear and comprehensive vision of provision and a well worked out set of priorities.

Competition for projects should be fair, transparent and must operate from a level playing field. The implementation of new appropriate technology should be encouraged and innovation should be employed to improve and develop the efficiency of procurement methods. Project proposals should be founded on sound financial analysis covering, where required, the whole life cycle and steps should be taken to assess the viability of alternatives. Every effort should be made to minimise the cost of bidding and protracted periods of competition should be avoided where possible.

Public provision of infrastructure finance

It is usual for government to invest in public infrastructure by means of raising finance from monies accrued from taxation. Taxation takes many different forms some of which will be directly associated with the infrastructure that they support. The most obvious example is the taxation of road users who may be required to pay fuel tax, vehicle licencing, road tolls and vehicle tax. In this way it could be argued that a national road network is self-financing, but in practice many countries succumb to the temptation to use this tax revenue to balance other public expenditure budgets. Infrastructure may be procured and managed throughout its life cycle by a national body, wholly owned and controlled by government, or by local government.

Governments commonly allow their expenditure to exceed revenue collected under taxation, the difference being covered by borrowing. The annual amount needed to cover the difference between spending and tax revenue is the public sector borrowing requirement (PSBR). The accumulated total of past government debt is known as the National Debt. Most governments deliberately operate in deficit, the amount of which depends on confidence in the future, when money borrowed will generate sufficient income to pay interest charges and reduce the total amount borrowed. Government borrowing is usually financed in the short term by selling Treasury Bills and in the long term by issuing Government Bonds. As a rule PSBR rises in recession and falls in boom periods. Where a government sells more bonds it must offer attractive interest rates, which in turn reduces the number of private bonds offered for sale and therefore private investment is held back. There are two other sources of public finance, namely National Savings schemes by individuals and by simply printing more money, the latter if allowed to get out of control will directly impact on the level of inflation.

Monetary policy therefore has a direct impact on how much can be spent on public infrastructure projects in any single year. As a general rule there will never be enough public money available to fund all infrastructure needs, hence projects may be cancelled or deferred and maintenance costs cut back to essential and affordable levels. This situation over time usually results in falling standards of service to users and may in some cases have the potential to constrain economic growth and development. This is particularly so where infrastructure provided is free to users. When undertaking a cost–benefit analysis care should be taken to identify latent benefits that may accrue to different aspects of national activity that may only become readily apparent when a particular infrastructure facility ceases to function.

Total government control over national monopolies, e.g. power generation, water supply, roads and railways through nationalisation has been argued to be in the national interest. At the centre of this argument is the notion of regulated prices and standards of service that are applied wholly in the public interest, including health and safety. Opponents of public ownership propose the counter argument that large organisations funded from the public purse lack the incentive to develop and improve, instead they suffer from bureaucracy and multiple subordination that creates inefficiency by soaking up funds unnecessarily.

It can therefore be argued that where infrastructure provision is strategically vital to the national interest, including public health and safety, then serious consideration should be given to wholly public ownership. This would be especially the case where it is difficult and costly to collect income from beneficiaries who indirectly benefit from the service or facility provided.

Private participation in wholly publicly owned infrastructure will normally take the form of the supply of expertise, materials and equipment in the design, construction and operation of specific projects. Public accountability is a key factor that controls the manner in which projects are procured and expenditure is controlled. Reference should be made to the procurement methods described in Chapter 6.

Private provision of infrastructure finance

Government policy will determine the extent and the potential for the provision of infrastructure by the private sector. It may be government policy to initiate private investment by participation in the purchase a whole national industry, e.g. the railway network or a regional authority that provides services such as water and electricity. The most obvious way to achieve this is by floating the organisation on the stock market. In the case of individual projects it is more likely that this will be achieved by a government initiated PPP involving some form of concessionary arrangement as described in the previous chapter. On the other hand, it is possible for an infrastructure proposal to be initiated by private investors who will seek the necessary approvals and conform with statutory requirements prior to implementation (PFI BOT Promotion, 1996).

An essential prerequisite of any project or service under consideration for private participation is the need to generate sufficient income over its life cycle, or concessionary period, to cover the cost of capital investment and operation with sufficient left over after all expenses have been paid to provide an acceptable return to investors (Walker and Smith, 1995).

Funds will be raised through one or more sources, the simplest being finance from a single individual or company. Usually infrastructure projects require large-scale investment and it is therefore more likely that there will be a consortium of investors, who may seek to combine their investment with borrowing from financial institutions and banks. An alternative would be to raise equity in the form of subscription to shares in a special purpose vehicle (SPV). Robust feasibility and investment studies linked to risk analysis will feature in the preliminary stages to assess need and the likely demand that will generate revenue income. Alternative scenarios should be explored and evaluated to optimise the best project solution.

Wholly private investment in public infrastructure does not involve the expenditure of public funds, which may result in government spending being commensurately reduced. Opponents of privatisation argue that in cases where the public is required to pay for infrastructure services, then this is potentially socially divisive in that the poorest in the community cannot afford the service provided and therefore only the better off members of society stand to benefit. Conversely, reduced public expenditure is known to benefit the national economy, which in turn provides scope and incentive for investment in new enterprises and ventures leading to higher employment and consumer spending. In the 1970s, the Thatcher government in UK realised the advantages to be gained by selling off public sector industries to private investors through flotation on the London Stock Exchange. This policy proved to be highly successful and resulted in the privatisation of most public services, including telecommunications, power generation and distribution, gas, water and transport. Private investment in infrastructure projects was also encouraged through the introduction of the private finance initiative (PFI). These steps helped to transform UK from a sluggish post war economy in the 1950s to the fourth largest economy in the world. This policy has been adapted and applied by most countries across the globe as a means of strengthening national economies and improving the potential to generate wealth and well-being.

Supporters of privatisation point to liberalisation and incentive to make profit, which when combined provide efficiency and better value for money services and products. By throwing off interference from government, decisions can be made based on business imperatives directed at the needs of the market, rather than politically motivated ideologies. The impact on public spending potentially reduces PSBR and reduces risk for government.

The privatisation of public companies and monopolies does not absolve government of fundamental responsibilities concerning public interest and safety. It is therefore essential for government to retain control over price, quality and regulations that govern health and safety (Smith, 1999). The overriding principle must require that the activities of privatised companies and monopolies offering essential services should not act against the public interest. Regulation will therefore be set in place at the point of privatisation. This may concern franchise agreements that will be periodically subject to new bids or renegotiation where performance benchmarks and safety standards will be reviewed. A watchdog is normally appointed to monitor and control performance and the regulation of prices. Usually an industry regulator will act in conjunction with government to oversee the activities of a privatised industry.

Typically a regulator appointed to oversee a service provided by a privatised company will perform the following functions:

- set price limits;
- ensure that prices are fair and reflect a fair profit;
- improved standards by means of greater efficiency;
- protection of customers by ensuring compliance with agreed responsibilities; and
- enforcement of fines and compensation payments where performance falls below agreed standards.

The regulator is further charged to ensure that companies are capable of adequately financing their function and that they are receiving a fair return for their investment. It is also important to establish that there are no undue

151

preferences between customers or discrimination in the way that companies fix and recover charges.

Public–private partnerships (PPP)

PPP are effectively applied to deliver infrastructure projects that would otherwise be out of the reach of available public funds (Levy, 1996). The market for PPP covers a wide range of projects ranging from bridges and tunnels to schools and hospitals. Income will be generated, either from a publicly-funded budget or from user charges, or a combination of both. The revenue generated over an agreed period of the concession should cover the capital and operational costs leaving a residual amount to provide an adequate return on investment (ROI), given economic conditions and the degree of risk involved. Where insufficient user income is generated then a case will be required for subsidies from the public purse to make up the shortfall (Akintoye et al., 2003).

PPP normally spans over a concessionary period, which may typically be between 10 and 30 years. Investment is mainly long term and will require very careful consideration of likely future charges and the degree of risk involved. Bids will be invited from private consortiums, who will be expected to prepare a proposal to design and construct the project and then to operate it over the concessionary period in accordance with agreed performance targets and quality standards. Working up the proposal is a time-consuming and costly business, hence bidding risk needs to be made reasonable by selecting a limited number of bidders, typically 3–4. It may be desirable to contribute to or pay the costs of the unsuccessful bidders. After a rigorous selection procedure a preferred bidder will be nominated and a process of negotiation will be commenced to develop and formalise an agreement. At this stage the successful consortium will be required to establish a special purpose vehicle (SPV) or to nominate an established company to run the project as a concessionaire. It is normal for the concessionaire to invest equity to cover a substantial proportion of the total project cost from its own resources (CIC, 1998). Borrowing from a variety of sources will provide the remainder of funding. Borrowing will be either secured on balance sheet against the asset bases of individual consortium members or it will be secured against cash flows generated by the project, which in turn will require controls to manage the risk involved. A typical example of a PPP project is shown in Figure 7.1.

A vital task will be to establish the project cash flow since the project will require regular substantial cash injections over the design and construction periods when there will be no revenue income. The length of the design and construct period is therefore crucial to the amount of interest to be paid on loans. Hence every effort should be made to expedite project handover after which revenue income can be generated to reduce the

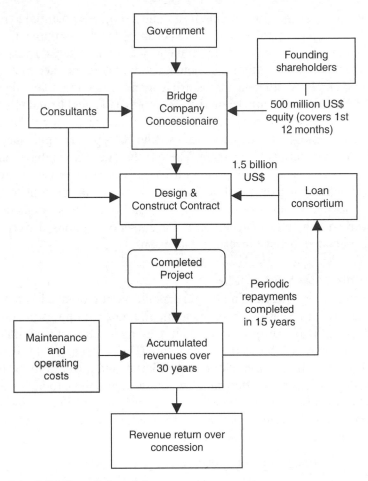

Figure 7.1   PPP financial model.

borrowing requirement. The optimisation of maintenance and operational costs will assist the achievement of the project breakeven point after which the project will begin to show a positive return. Close control should also be maintained over the achievement of agreed performance benchmarks to avoid incurring unnecessary cost penalties. Details of methods and procedures are covered in the monitoring and control section of the chapter.

Project financial analysis

Every project will have different types of costs, revenues and risks associated with its development, implementation and operation. Because of the scale of infrastructure projects and the risk often associated with long-term investment, it is important that the financial feasibility of a project is fully investigated. There are various techniques available to evaluate the financial viability of projects in terms of the structure of costs, revenues and associated risk issues. The amount of risk and a project's potential to realise its life cycle objectives, given anticipated changes to its environment, should be evaluated using the best and most reliable data. Other likely developments such as technology breakthrough should also be taken into account. There will also be areas where data are insufficient and where intuitive judgement is required to assess the impact on risk. To improve the quality and reliability of investment decisions about the provision of infrastructure it will be necessary to use a range of techniques and methods to provide guidance as a basis for more informed judgement. These techniques, in combination, can be complimentary and may improve confidence, on the contrary, where they provide contradicting results they may only serve to increase uncertainty, thereby making the investor more cautious. The following sections provide fully worked examples of techniques that provide solutions derived from analysis intended to assist investment decisions.

Discounted cash flow (DCF)

This technique is for assessing the return on capital employed in a project over a suitable time period or its economic life. Discounting is the theoretical adjustment for the time value of money as it looks at how much an investment today will be worth in the future (Damoderan, 2001). This will depend on one's perception of demand, inflation and uncertainty. The fundamental question can be illustrated by asking what will be today's value of £1 000 000 received in 10-years time. The reverse question is how much would need to be invested today to yield £1 000 000 in 10 years and this can be determined by applying compound interest, i.e.

$$\text{Future value of cash flow } (CFt) = CFo\,(1 + r)^t$$

where CFo = present value of cash flow

r = interest

t = number of years

Hence, £1 000 000 = CFo $(1 + r)^t$ where $r = 10\%$ and $t = 10$ years,

$$CFo = \frac{£1\,000\,000}{(1.10)^t} = £385\,544$$

Thus £1 000 000 received in 10 years time assuming a discount rate of 10% will be the equivalent value of £385 544 received in cash today. If one could accurately predict future interest rates over the upcoming 10 years then the above answer would hold good under any circumstances. However, it is highly unlikely that interest rates will remain the same, due primarily to uncertainties previously mentioned. Thus the selected interest rate reflects how one sees the future and the uncertainties associated with a particular project.

This principle can be usefully applied by estimating annual cash flow for each year of the entire project, comprising of capital expenses (CE), operating expenses (OE) and revenue income (RI). Since resources expended and income gained will be accruing annually, each year's cash flow will need to be adjusted according to the discount factor for that year (see discount table in Appendix 1).

Assuming an infrastructure project to have an estimated CE of £98.5 million over a design and construct period of 5 years, this expenditure can be expressed as a series of annual budgets. Similarly, the operating period of 9 years can be expressed in the same manner using OE and RI, thus producing RI.

It is essential that correct decisions are taken at the design stage when the cost of making changes is relatively small and the potential to make improvements is high. Therefore all possible options should be considered and the best possible solution should be selected that meets the client requirements and provides optimal operational performance.

Table 7.1 illustrates a comparison between two design options where non-cumulative and cumulative annual DCF is taken throughout the period of the whole project. The calculation does not take into account the cost of borrowing, which should be dealt with elsewhere, nor does it allow for the deduction of tax on revenue. The discount rate is set at 10%. Design A generates a cumulative net present value (NPV) of £96.2465 million, thus the project returns a shortfall of £2.2535 million compared with the CE of £98.5 million. Design B incurs additional capital cost, however operating expenditure is less and there is the added benefit of the design and construction period being reduced to 4 years, thereby providing an additional year's income. Taking these factors into account, the cumulative NPV increases to £122.1648 million which exceeds the £105 million capital expenditure by £17.1648 million.

Assuming that there are no other factors to be taken into account then it is reasonable to conclude that Design B offers a better proposition, given that it produces a further £17.1648 million above the NPV discounted at 10%.

Sensitivity analysis

Sensitivity analysis is used when accurate probabilistic information is not available and uncertainty surrounds project data concerning returns, costs and the selected discount rate. Once the NPV has been calculated, then one aspect of the original data can be selected for analysis, e.g. the discount rate. Taking Design B as the preferred option four discount values have been analysed according to the following equation:

$$\text{NPVproj} = \sum_{t=1}^{t=10} \text{CF}t \left(\frac{1}{1 + r^t} \right)$$

Hence,

NPVproj at
5% = +£187.7997 millions
10% = +£122.1648 millions
15% = +£80.9341 millions
20% = +£57.001 millions

Table 7.1 Infrastructure designs A and B

	Design and build						Operate								
Year	0	1	2	3	4	5	6	7	8	9	10	11	12	13	14
DESIGN A															
Capital expense (CE)	(−98.5)														
−Cash flow															
Operating expense (OE)		(−5)	(−4.5)	(−26)	(−35)	(−28)	−15	−15	−15	−15	−15	−15	−15	−15	−15
Revenue income (RI)							30	40	45	45	45	45	45	45	45
Gross revenue[a]							15	25	30	30	30	30	30	30	30
Discount rate 10%	1	0.9091	0.8264	0.7513	0.683	0.6209	0.5645	0.5132	0.4665	0.4241	0.3856	0.3505	0.3186	0.2897	0.2633
NPV							8.4675	12.83	13.995	12.723	11.568	10.515	9.558	8.691	7.899
CUM NPV							8.4675	21.2975	35.2925	48.0155	59.5835	70.0985	79.6565	88.3475	96.2465
Project capital expense (CE)	£98.5 million														
Total operating expenses (OE)	£135 million														
Total revenue income (RI)	£385 million														

156

157

Discount rate: 10%

Project NPV shortfall: Shortfall £2.2535 million

DESIGN B

Capital expense (CE)	(−105)														
−Cash flow		(−7)	(−18)	(−45)	(−35)										
Operating expense (OE)						−13	−13	−13	−13	−13	−13	−13	−13	−13	−13
Revenue income (RI)						30	40	45	45	45	45	45	45	45	45
Gross revenue[a]						17	27	32	32	32	32	32	32	32	32
Discount rate 10%	1	0.9091	0.8264	0.7513	0.683	0.6209	0.5645	0.5132	0.4665	0.4241	0.3856	0.3505	0.3186	0.2897	0.2633
NPV						10.5553	15.2415	16.4224	14.928	13.5712	12.3392	11.216	10.1952	9.2704	8.4256
CUM NPV						10.5553	25.7968	42.2192	57.1472	70.7184	83.0576	94.2736	104.4688	113.7392	122.1648

[a] Tax liability not deducted

Project capital expense (CE): £105 million

Total operating expenses (OE): £130 million

Total revenue income (RI): £430 million

Discount rate: 10%

Project exceeds NPV by: £17.1648 million

Graph 7.1 shows these results plotted to describe a curve indicating the relationship between NPVproj and the discount rate. The outcome demonstrates that the project has strong viability given that zero NPVproject produces a discount rate in the region of 11.75%. The curve also illustrates that less risky projects carrying a lower discount rate show a greater NPV. The eventual outcome of the project will be somewhat different according to prevailing circumstances, hence given the same income, a lower discounted project would be more profitable, but the probability of this occurring would be potentially more remote.

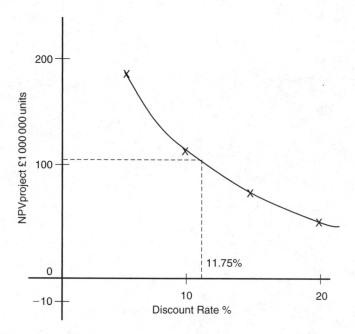

Graph 7.1   Project B: NPVproject relationship with discount rate r.

Break-even analysis

The volume of revenue largely determines when or if a project is capable of showing a ROI. This may be expressed simply as three elements in the case of Design B, i.e.

1. Capital expense (CE)
2. Operating expense (OE)
3. Revenue income (RI)

Figure 7.2 shows an early break-even point occurring at year 3.7; however this neither takes in account of the cost of borrowing, nor does it take any account of any adjustment for NPV. Therefore simple break-even analysis in the context of project appraisal is only indicative and is of limited use.

158

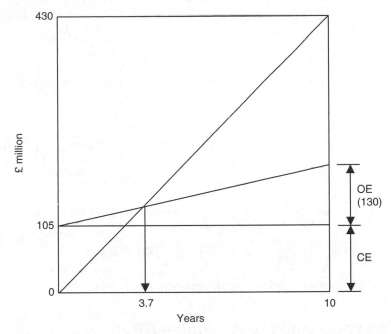

Figure 7.2   Break-even analysis: Design B.

Break-even analysis can be usefully extended to determine the point at which NPV switches from positive to negative. This can be calculated by taking into account capital depreciation and corporation tax as follows:

(a) Annual project costs £13 000 000

(b) Depreciation (D) based on capital
Investment £105 million depreciated
over 10 years by straight line £10 500 000

Pre-tax profit = [Revenue (R) – Costs(a + b)]
After tax profit @ 40% = 0.6(R – 23.5)
Net annual cash flow   = D + 0.6(R − 23.5)
$$10.5 + 0.6(R − 23.5)$$

Taking the annuity factor for 10% over 10 years as 6.145 (see Table A2 in Appendix)

$$PV = 6.145 \times (10.5 + 0.60(R − 23.5))$$

To determine the break-even point let PV = investment. Hence,

$$6.145 \times (10.5 + 0.60(R − 23.5)) = 105$$
$$3.687R − 22.122 = 105$$
$$R = 34.478 \ (000)$$

Therefore the annual revenue required to produce a zero–PV at 10% over 10 years is £34.478 million. This is graphically expressed as in Figure 7.3.

159

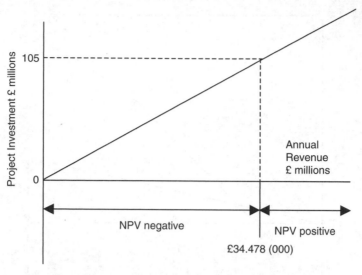

Figure 7.3   Annual revenue required to produce zero NPV at 10%.

Using uncertainty to determine NPVs

Uncertainty can be taken into account by articulating a range of outcomes from the most optimistic to the most pessimistic (Samuels and Wilkes, 1986). Taking Design B as an example the investment equates to the total cost of the completed project set at £105 million and this will set as year 0.

Project statistics:

(a) Total annual average revenue (Years 5–14)	£43 000 000
(b) Total annual expenditure (Years 5–14)	£13 000 000
(c) Depreciation (Straight line)	
105 000 000/10 years	£10 500 000
Pretax profit (a − b − c)	**£19 500 000**
Less taxation @ 40%	£7 800 000
Profit after tax	**£11 700 000**
Cash flow from operations:	
10 500 000 + 11 700 000	£22 200 000

Assuming a discount rate of 10%

Present Value (PV) = Annual cash flow × 10 Year Annuity Factor (See Table A2 Appendix: 6.145.)

$$PV = 22\,200\,000 \times 6.145 = £136\,419\,000$$
$$NPV = PV - \text{Investment}$$
$$= £136\,419\,000 - £105\,000\,000$$
$$= £31\,419\,000$$

The left-hand side of Table 7.2 shows the expected, optimistic and pessimistic values based on an analysis of probability or intuitive judgement based on

160

Table 7.2   NPV uncertainty

Variable	Range (£ million)			NPV (£ million)		
	Pessimistic	Expected	Optimistic	Pessimistic	Expected	Optimistic
Investment	115	105	95	23.887	31.419	38.961
Revenue	40	43	45	20.358	31.419	42.48
Costs	14	13	12	27.732	31.419	35.106

all available facts. The right-hand side of the table shows NPV if the variables are set individually to their respective optimistic and pessimistic values.

The costs are adjusted as follows:

[13 (expected) − 12 (optimistic)] × (1 − tax @ 40%)
Hence, 1 × 0.6 = £0.6 million
0.6 × 6.145 (10 year annuity factor) = £3.687 million

Optimistic: 31.419 + 3.687 = 35.106
Pessimistic: 31.419 − 3.687 = 27.732

Investment is adjusted as follows:

Optimistic

a. Revenue	£43 000 000
b. Costs	£13 000 000
c. Depreciation (£95 millions ÷ 10)	£9 500 000
d. Pretax profit (a − b − c)	£20 500 000
e. Taxes at 40%	£8 200 000
Profit after tax (d − e)	**£12 300 000**
f. Cash flow (c + e)	£21 800 000

PV = £21.8 million × 10 year annuity at 10%
   = 21.8 × 6.145
   = £133.961 million
NPVoptimistic = PV − Investment opt
   = 133.961 − 95
   = £38.961 million

Pessimistic

a. Revenue	£43 000 000
b. Costs	£13 000 000
c. Depreciation (£115 millions ÷ 10)	£11 500 000
d. Pretax profit (a − b − c)	£18 500 000
e. Taxes at 40%	£7 400 000
Profit after tax (d − e)	**11 100 000**
f. Cash flow (c + e)	£22 600 000

$$PV = £22.6 \text{ million} \times 10 \text{ year annuity @ } 10\%$$
$$= 22.6 \times 6.145$$
$$= £138.877 \text{ million}$$

NPVpessimistic $= PV - \text{Investment opt}$
$$= 138.877 - 115$$
$$= £23.877 \text{ million}$$

Revenue is adjusted as follows:

[43 (expected) − 40 (pessimistic)] × (1 − Taxation @ 40%)

[46 (optimistic) − 43 (expected)] × (1 − Taxation @ 40%)

Hence, $3 \times 0.6 = £1.8$ million

$PV = 1.8 \times 6.145 = £11.061$

NPVoptimistic $= 31.419 + 11.061 = £42.480$ million
NPVpessimistic $= 31.419 - 11.061 = £20.358$ million

The outcome of this analysis shows that the project is a strong investment and will still show good profits under the most pessimistically perceived circumstances.

Project risk analysis

This section describes how probabilistic data can be used in the evaluation of investment in infrastructure projects under conditions of risk. It is usual that risk is associated with probability is normally distributed about the mean value χ and that variance can be measured in terms of standard deviation δ. Under these circumstances the parameters concerning the distribution of probability are known. In cases where uncertainty prevails then techniques associated with decision-making under uncertainty should be considered.

Taking the case of a single project where profitability is considered on an annual basis, the probability of returns over a 3 year period is shown below:

Distribution of probable returns ('000s)

Year 1		Year 2		Year 3	
Return	Probability (P)	Return	Probability (P)	Return	Probability (P)
12	0.1	18	0.4	11	0.3
15	0.5	27	0.6	15	0.4
18	0.4			21	0.3

The expected return χ for Year 1 is calculated as follows:

X_{ij}	P_{ij}	$P_{ij}X_{ij}$
12	0.1	1.2
15	0.5	7.5
18	0.4	7.2

162

Therefore: $\sum_{i=1}^{3} P_{ij}X_{ij} = \chi_1 = 15.9$

Similar calculations for years 2 and 3 give the following:

$\chi_2 = 23.4$
$\chi_3 = 15.6$

Hence the gross NPV calculated at 10% discount rate is

Year 1: $15.9 \times 0.9091 = 14.455$
Year 2: $23.4 \times 0.8264 = 19.338$
Year 3: $15.6 \times 0.7513 = 11.720$
Expected gross PV $= 45.513$

Assuming the outlay for the 3-year project is £43 000 then the expected net present return (ENPR) will amount to

Expected gross PV − Initial outlay
Therefore ENPR = 45.513 − 43 = £2.513 K.

It may be that more is known about the distribution of probability from year to year. Table 7.3 sets out the combinations of cash flow over the 3 years and the probability of such an occurrence. From each set of yearly results

Table 7.3   ENPV calculation

Cash flow ('000s) Years			Probability P	NPV 10%	$P * $ NPV
1	2	3			
12	18	11	0.0110	−8.9513	−0.098
12	18	15	0.0210	−5.9461	−0.125
12	18	21	0.0090	−1.4383	−0.013
12	27	11	0.0170	−1.5137	−0.026
12	27	15	0.0290	1.4915	0.043
12	27	21	0.0130	5.9993	0.078
15	18	11	0.0700	−6.2240	−0.436
15	18	15	0.1220	−3.2188	−0.393
15	18	21	0.0470	1.2890	0.061
15	27	11	0.1090	1.2136	0.132
15	27	15	0.1810	4.2188	0.764
15	27	21	0.0710	8.7266	0.620
18	18	11	0.0370	−3.4967	−0.129
18	18	15	0.0590	−0.4915	−0.029
18	18	21	0.0250	4.0163	0.100
18	27	11	0.0530	3.9409	0.209
18	27	15	0.0910	6.9461	0.632
18	27	21	0.0350	11.4539	0.401
			1.0000	ENPV	1.791

the NPV is calculated at 10% discount and the sum of $p \times$ NPV provides the Expected Net Present Value (ENPV), e.g. the first row of figures shown in Table 7.3 is calculated as follows:

$$
\begin{aligned}
\text{Cash flow:} \quad 12 \times 0.9091 &= 10.9092 \\
18 \times 0.8264 &= 14.8752 \\
11 \times 0.7513 &= 8.2643 \\
\text{Total} \quad & 34.0487 \\
\text{Initial outlay} \quad & 43.0000 \\
\textbf{NPV} &= \mathbf{-8.9513}
\end{aligned}
$$

Subsequent rows are calculated in the same manner.

It will be noted that there is a difference between ENPV compared to the ENPR of the previous example. This is because of greater information being available about the distribution of probability.

The standard deviation associated with NPV can be calculated. Hence variance represented by the square of the standard deviation is the average squared departure of NPV from its norm value. This is a measure of total risk for the investment.

$$\delta\text{NPV} = \sqrt{\Sigma P(\text{NPV} - \text{ENPV})^2}$$

Probability (P)	NPV 10%	NPV $-$ ENPV	(NPV $-$ ENPV)^2	P (NPV $-$ ENPV)^2
0.0110	-8.9513	-10.7421	115.393	1.269
0.0210	-5.9461	-7.7369	59.860	1.257
0.0090	-1.4383	-3.2291	10.427	0.094
0.0170	-1.5137	-3.3045	10.920	0.186
0.0290	1.4915	-0.2993	0.090	0.003
0.0130	5.9993	4.2085	17.711	0.230
0.0700	-6.2240	-8.0148	64.237	4.497
0.1220	-3.2188	-5.0096	25.096	3.062
0.0470	1.2890	-0.5018	0.252	0.012
0.1090	1.2136	-0.5772	0.333	0.036
0.1810	4.2188	2.4280	5.895	1.067
0.0710	8.7266	6.9358	48.105	3.415
0.0370	-3.4967	-5.2875	27.958	1.034
0.0590	-0.4915	-2.2823	5.209	0.307
0.0250	4.0163	2.2255	4.953	0.124
0.0530	3.9409	2.1501	4.623	0.245
0.0910	6.9461	5.1553	26.577	2.419
0.0350	11.4539	9.6631	93.376	3.268
				22.525

Standard Deviation $\delta = \sqrt{22.525} \times £1000 = £4746$

The coefficient of variance (*cov*) profiles a measure of risk in that the lower the value provided by dividing the standard deviation by ENPV, the less the risk involved, hence

$$cov = \frac{4.746}{1.791} = 2.65$$

From the example NPV = 0 at ENPV/standard deviation δ. Hence, (1.791/4.746) = 0.377 standard deviation below the mean.

It is now necessary to determine whether *cov* = 2.65 represents high or low risk. Assuming that the distribution of NPV is normal then the area under the curve corresponds to 1.96δ or 95% of the observations of the *x* variable lie within this range. By reference to area under the normal density function Table A3 in Appendix.

0.37   0.6443   difference is 0.0037 × 0.7 = 0.00259 + 0.6443

0.38   0.6480   0.6443 + 0.00259 = 0.64689 (64.7%)

Hence, there is a 35.3% chance of making a loss below the 10% criterion discount rate. This may be perceived too high a risk. To achieve an arbitrary rate of say 10% the result should be 1.28 standard deviations above the mean. This implies that the *cov* should be no greater than (1/1.28) = 0.78. Thus the *cov* of 2.65 is too high and the project would seem to be carrying a high degree of risk.

Inflation

When considering major, medium or long-term infrastructure investment decisions it is important to account for uncertainty caused by inflation. The crucial issue concerns what adjustments should be made to the DCF calculations and the extent to which the project remains viable.

Taking annual revenue (*r*) and costs (*c*) generated by Design B shown in Table 7.1. It has been estimated that the average annual rate of cost inflation from inception of the project will be at 10%, while revenue inflation will be restricted to 5% per annum. It will therefore be necessary to make adjustments to the annual operating income as follows:

Annual revenue $(r)t = r*(1.05)^t$ and likewise annual cost $(c)t = c*(1.10)^t$

Annual excess of revenue over cost *Rt* is generated by $rt - ct$ and a further adjustment is made according to the discount rate (i.e. 10%).

Years (*t*)	Revenue (*r*)	*rt*	Cost (*c*)	*ct*	Rt (*rt − ct*)	PV at 10%	Adjusted Rt (£)
5	30	38.29	13	20.94	17.35	0.6209	10.774
6	40	53.60	13	23.03	30.57	0.5645	17.259
7	45	63.32	13	25.33	37.99	0.5132	19.495

165

Years (t)	Revenue (r)	rt	Cost (c)	ct	Rt (rt − ct)	PV at 10%	Adjusted Rt (£)
8	45	66.49	13	27.87	38.62	0.4665	18.016
9	45	69.81	13	30.65	39.16	0.4241	16.606
10	45	73.30	13	33.72	39.58	0.3855	15.259
11	45	76.97	13	37.09	39.87	0.3505	13.976
12	45	80.81	13	40.80	40.01	0.3186	12.748
13	45	84.85	13	44.88	39.97	0.2897	11.581
14	45	89.10	13	49.37	39.73	0.2633	10.461
					Adjusted NPV		146.174

Because RI is well in excess of costs NPV is increased in value despite cost inflation being double. Hence NPVproject has been increased by £24.009 million.

In the event that revenue inflation is set at zero, while average costs increase at 10% per annum the adjusted NPVproject would amount to only £47.048 million. Under these circumstances the project would potentially generate a loss of £57.95 million.

Pricing decisions are also related to inflation and it is necessary to make appropriate provision. The following example illustrates the case of a water company and the requirement to supply a large industrial customer. Over the next 3 years customer requirements are estimated to be 120, 160 and 200 million litres respectively. Cost inflation is judged to be at 5% per year and the cost of capital is 10%.

There are two techniques available for supplying the water involving the following costs:

Years	Technique 1	Technique 2
1	$4 million	$6 million
2	$5 million	$5 million
3	$6.5 million	$5 million

It is required that a break-even point on NPV should be reached at the end of the 3-year period. The charge for water will be fixed at the same rate over the 3-year period and no other charges will be incurred.

Solving for the price per litre of water supplied P1:

Technique 1

$$NPV1 = \{[-4(1.05) + 120P1]1.1 - 5(1.05)^2 + 160P1\}1.1 - 6.5(1.05)^3 + 200P1 = 0$$

Therefore P1 = $0.036 per litre

Technique 2

$$NPV2 = \{[-6(1.05) + 120P1]1.1 - 5(1.05)^2 + 160P1\}1.1$$
$$- 5(1.05)^3 + 200P1 = 0$$

Therefore P2 = $0.0374 per litre

Hence Technique 1 offers a cheaper solution.

Cost of borrowing

The cost of borrowing to support one or more infrastructure projects will depend on a well-developed proposal that provides adequate revenues to generate an adequate return with an acceptable level of risk. Socio-economic and political stability provided by the context within which the project(s) is set will have a direct bearing on the perceived risk involved. Geographical and environmental factors will also need to be taken into account.

Longer lending terms will increase the degree of uncertainty and it is therefore essential that the lenders are able to identify a reliable and steady income stream that results from the project(s). Where higher levels of risk are identified, lenders will require higher returns on loans made and they may even seek additional returns, e.g. royalties. In the case of very large projects it is usual to establish banking syndicates to provide sufficient lending and to spread the risk involved.

Lending facilities will normally be provided from one or more banks usually in the form of a loan with specific conditions attached covering interest charges and arrangements for repayment. Such loans will at one extreme carry all the risk while at the other they may be totally secured against bonds to guarantee repayment in full in the event of default.

Where loans are made for infrastructure projects it is likely that they will be secured against the asset value and income generated by the project. This means that in the case of default, the lenders will take control of the project. Fixed term loans can be either repayable at the end of the loan period or repayable in periodic instalments.

Where borrowing is judged to be short term it may be possible to negotiate an overdraft facility. This will depend on the credit worthiness of the borrower, which will also affect the level of interest charged. This form of borrowing has the advantage that interest is charged at the prevailing bank rate, plus an additional percentage to cover risk and other special considerations. Normally, this is a very cheap way of borrowing compared with fixed-term loans where interest is likely to be compounded based on the full amount of the loan, rather than the actual sum outstanding on a month by month basis. The main disadvantage with overdraft facilities is that interest rates may rise and the loan can be called in at any time by the bank. Where stable low-risk conditions prevail, the bank might be amenable to extending overdraft arrangements indefinitely, thus providing a form of cheap medium-term finance.

With reference to Design B depicted in Table 7.1 it is possible to calculate the cost of borrowing to support the project. This will help to provide a better idea of the real return within the global NPV calculation. Much will depend on the final conditions negotiated with the lender and the degree of flexibility available to restructure the condition of the loan according to changing needs and circumstances.

Taking all matters into consideration, including equity provided from internal sources, it may be necessary to take out a £70 million loan at say 7% compound interest over a fixed term of 5 years. The total cost of which would be

$$70(1.07)^{05} = 98.179 - 70.0 = £28.179 \text{ million}$$

Assuming the loan is taken out at year 3 and is repaid in full at year 8 then the NPV of the loan at 10% DCF is 0.4665, therefore the NPV project deduction will be £28.179 × 0.4665 = £13.146 million. In other words, the loan is a good deal since it will free up resources to invest in other projects, especially taking into consideration the criterion rate of discount set at 10% discount. Any short-term borrowing requirement could be covered by overdraft arrangements agreed with the bank or lending consortium.

In the case of contractors undertaking infrastructure projects overseas it is possible to apply for 'export credit guarantees' from a relevant government department. This is not a loan, but by offering to guarantee payment in the event of default of an overseas customer, banks are more likely to offer loans at reasonable rates of interest. In this manner governments seek to promote overseas trade alongside other concessionary loan arrangements.

Performance bonds are awarded to companies that take on large overseas infrastructure projects that involve more than the worth of company or consortium assets. Normally the bank will issue a performance bond to a foreign customer who will only ask for the bond to be met in the event of default. This will carry a cost in addition to interest payments.

Stochastic decision trees

Stochastic decision trees are characterised by the incorporation of chance node (CN) and decision nodes (DN), thus producing a multi-stage process within an overall decision-making framework (Waters, 1989). Decision-making nodes are identified by a square box that requires a conscious decision to select one of the branches that emanate from the node. A CN is designated by a circle that shows different states, actions or conditions that are presented for a chosen strategy. Normally there are two types of branches that connect the nodes. Decision branches are denoted by parallel connecting lines representing a strategic course of action, whereas chance branches indicate probability associated with state, action or condition.

THE FINANCING OF INFRASTRUCTURE PROJECTS

Stochastic decision trees provide strategic planners with a framework for the consideration of probabilistic multi-stage decision processes and they are ideally suited to the evaluation of alternative strategies for capital projects.

It is possible that there will be more than one strategy under consideration for the provision of infrastructure. Further, the analysis can be extended to usefully incorporate ENPVs as a means of providing a consistent basis on which to make financial judgements.

Take the hypothetical case of building a large power generation plant to meet anticipated future demand, or alternatively the construction of a small plant that can be subsequently expanded if demand remains to be high.

Strategy 1

- Construct a large power generation plant at a cost of £500 million.
- It is estimated that high demand will produce a revenue of £125 million per annum over a 10-year period and low demand will be £75 million over the same period. The probability of high demand is estimated to be 0.70 and low demand is 0.30.

Strategy 2

- Construct a small power generation plant at a cost of £250 million.
- In the event that demand is high at £75 million per annum a decision must be taken to expand the plant at a cost of £350 million ready to commence generation at the end of year 3. The expanded plant is estimated to have a 0.70 chance of high demand at £150 million per annum for 7 years and a low demand of £75 million for the same period. In the event that the plant is not expanded it is estimated that there will be a 0.70 chance of high demand at £75 million per annum for 7 years and low is estimated to be 0.30 probability at £35 million per annum for the same period.
- It is estimated that there will be a 0.30 chance that demand will be low on the small power generation plant at £35 million per annum. In this case the plant will not be expanded.

The objective is to calculate the expected NPV for each strategy and to determine the most beneficial course of action.

Assumptions

1. The discount rate will be at 10%.
2. Existing power generation will not be interrupted by the expansion.
3. The expansion of the small power plant will incur the full cost at the end of year 3.

The first step is to construct the stochastic decision tree to represent the problem as follows:

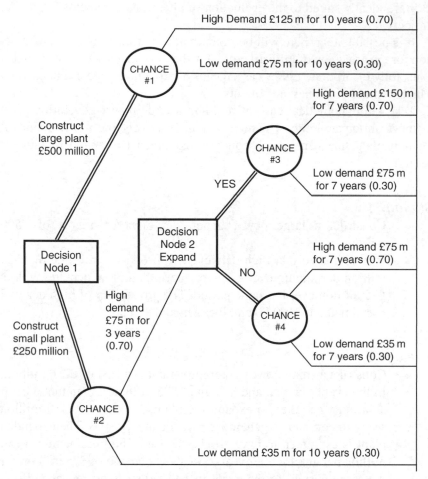

The analysis of the decision tree requires the calculation of values for CN and DN nodes. This involves a rollback procedure commencing at the right hand end of the decision tree and working backwards to the first DN (DN#1).

Chance Node CN#1

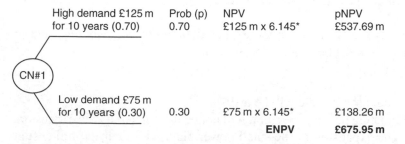

	Prob (p)	NPV	pNPV
High demand £125 m for 10 years (0.70)	0.70	£125 m x 6.145*	£537.69 m
Low demand £75 m for 10 years (0.30)	0.30	£75 m x 6.145*	£138.26 m
		ENPV	**£675.95 m**

*From Annuity table: PV of £1 invested for each of 10 years @ 10% discount rate = £6.145

170

Chance Node CN#3

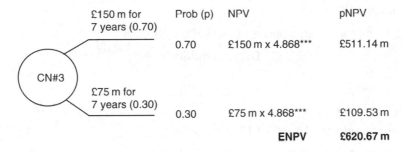

	Prob (p)	NPV	pNPV
£150 m for 7 years (0.70)	0.70	£150 m x 4.868***	£511.14 m
£75 m for 7 years (0.30)	0.30	£75 m x 4.868***	£109.53 m
		ENPV	**£620.67 m**

Chance Node CN#4

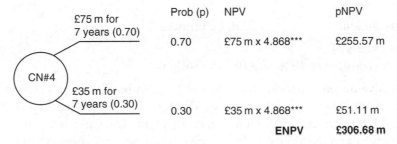

	Prob (p)	NPV	pNPV
£75 m for 7 years (0.70)	0.70	£75 m x 4.868***	£255.57 m
£35 m for 7 years (0.30)	0.30	£35 m x 4.868***	£51.11 m
		ENPV	**£306.68 m**

*** From Annuity table: PV of £1 invested for each of 7 years @ 10% discount = 4.4868

Decision Node #2

Rolling back to decision node #2 the highest value CN is selected as the most advantageous.

CN#3	£620.67 millions	CN#4	£306.68

Less

Plant expansion   £350.00 millions

CN#3 Net value   £270.67 millions

Hence the position value of DN#2 = £306.68

Chance Node #2

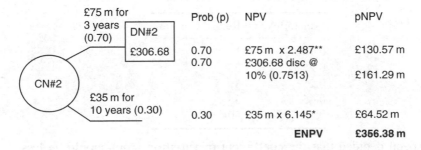

	Prob (p)	NPV	pNPV
£75 m for 3 years (0.70) DN#2 £306.68	0.70	£75 m x 2.487**	£130.57 m
	0.70	£306.68 disc @ 10% (0.7513)	£161.29 m
£35 m for 10 years (0.30)	0.30	£35 m x 6.145*	£64.52 m
		ENPV	**£356.38 m**

**From Annuity table: PV of £1 invested for each of 3 years = £2.487

Decision Node #1

Rolling back to DN#1

Strategy 1 Construct a large power plant (£500 millions)

CN#1	ENPV	£675.95 millions
Less		
Capital Investment		£500.00 millions

**Return in excess of
criterion discount rate 10%   £175.95 millions**

Strategy 2 Construct a small power plant (£250 millions)

CN#2	ENPV	£356.38 millions
Less		
Capital Investment		£250.00 millions

**Return in excess of
criterion discount rate 10%   £106.38 millions**

The most advantageous option based on ENPV is to build a large power generation plant, since this potentially yields £69.57 millions more when compared to the strategy of building a small plant. Decision trees are highly dependent on the ability to determine realistic strategic options and the judgement of probability.

Risk assessment using Monte–Carlo simulation

Monte–Carlo simulation has proved to be a relatively simple and straight forward technique for examining certain attributes or aspects of a project by user assessment of the likelihood of achieving certain goals. These are then allocated with random numbers as a means of simulating chance. By repeating the random number selection process many times a probability distribution is built up which provides an indication risk (Taha, 1982).

To illustrate the application of Monte–Carlo simulation a simple example is taken where the known cost of an infrastructure is £125 million. The project will earn revenues for 6 years and the criterion discount rate will be 10%. The risk associated with the project is represented by the probability of achieving the following distribution of returns:

Probability	Return £ million
0.10	15
0.25	25
0.25	30
0.25	35
0.15	50

It has been decided that the coefficient of variation (*cov*) should be less than 0.7.

172

Step 1

Two digit numbers are allocated to the probability range of each category of return, i.e.

Return (R)	Allocated No.
15	00–09
25	10–34
30	35–59
35	60–84
50	85–99

Step 2

The random number table contained in the appendix is then used to determine the selection of the simulated returns, e.g. No. 31 falls within the range 10–34, hence a return of £25 million is selected. This process is repeated for each of the 6 years to complete one simulation run. Any order of random numbers may be selected from the table provided that the same number is not used more than once. In this manner an unbiased simulation of chance is achieved.

Step 3

The following table is constructed showing six simulation runs. The results for each year are then discounted by year and the NPV is calculated for each simulation run.

Year	Discount	Simulation run					
	10%	1	2	3	4	5	6
0		−125	−125	−125	−125	−125	−125
1	0.9091	25	25	30	35	25	35
2	0.8264	35	35	35	35	30	35
3	0.7513	30	30	30	35	30	25
4	0.683	50	25	30	25	15	25
5	0.6209	25	50	25	15	35	35
6	0.5645	35	30	50	15	50	50
	NPV	18.6205	14.2455	17.9735	−3.106	5.26	21.5565

The average of all the NPV values is calculated to provide the ENPV, hence

```
+18.6205
+14.2455
+17.9735
 −3.1060
 +5.2600
+21.5565
+74.5500
```

Therefore, ENPV $= \dfrac{74.5500}{6} = 12.425$

The standard deviation of NPV is

$$\delta NPV = \sqrt{\frac{\Sigma(NPV - ENPV)^2}{n - 1}}$$

where n is the number of simulation runs.

NPV	ENPV	NPV − ENPV	$(NPV - ENPV)^2$
18.6205	12.425	6.1955	38.3842
14.2455	12.425	1.8205	3.3142
17.9735	12.425	5.5485	30.7859
−3.106	12.425	−15.531	241.2120
5.260	12.425	−7.165	51.3372
21.5565	12.425	9.1315	83.3843
	$\Sigma(NPV - ENPV)^2$		448.4178

Therefore $\delta NPV = \sqrt{448.42/5} = 9.47$

The $cov = \delta NPV/ENPV = 9.47/12.425 = 0.762$

The cov is very near to 0.7 and because of the limited number of runs the project should not be rejected at this stage. Many more runs should be carried out to provide a reliable distribution of results from which reasonable predictions can be made.

These results can be further checked by direct calculation. The mean yearly return is calculated from the table of probabilities and returns to be

Mean return $= 0.1(15) + 0.25(25) + 0.25(30) + 0.25(35) + 0.15(50)$
$= £31.5$ million

The variance of return in any year $\delta^2 R = \Sigma P(R - 31.5)^2$, where P and R are the values in the probability table in previous page, therefore

$0.10(15 - 31.5)^2$	27.225
$0.25(25 - 31.5)^2$	10.563
$0.25(30 - 31.5)^2$	0.563
$0.25(35 - 31.5)^2$	3.063
$0.15(50 - 31.5)^2$	51.338
Σ	92.752

The variance of NPV can be expressed as

$$\delta^2 NPV = \sum_{t=0}^{n} \left[\frac{\delta^2 t}{(1 + r)^{2t}} \right]$$

Hence using the above where $n = 6$,

$$\delta^2 \text{NPV} = \delta^2 0 + \frac{\delta^2 1}{(1+r)^{2t}} + \frac{\delta^2 2}{(1+r)^{2t}} \cdots \frac{\delta^2 6}{(1+r)^{12}}$$

Thus,

$\delta^2 \text{NPV}$

$$= 0 + 92.752 \left[\frac{1}{(1.1)^2} + \frac{1}{(1.1)^4} + \frac{1}{(1.1)^6} + \frac{1}{(1.1)^8} + \frac{1}{(1.1)^{10}} + \frac{1}{(1.1)^{12}} \right]$$

$$\delta \text{NPV} = \sqrt{92.752(3.245)} = 17.35$$

ENPV is now calculated by discounting the expected return in each year:

ENPV $= -125 + 31.5(4.3553) = 12.19$ (See Table 2: Present value of an annuity of £1.)

The true *cov* is calculated by dividing δNPV by ENPV, thus:

$$cov = 17.35/12.19 = 1.423$$

There is a significant difference between the theoretical true *cov* of 1.413 and the simulated *cov* of 0.762, which at this stage simply confirms that more simulation runs are required. Furthermore

NPV $= 0$ at $12.19/17.35 = 0.703$ standard deviations below the mean.

Therefore from the Normal distribution Table 3:

0.70   0.7580   difference is $0.0031 \times 0.3 = 0.00093$
0.71   0.7611 $0.7580 + 0.0009 = 0.7589$

Thus at this stage of the simulation there is a 24.11% chance of making a loss below the 10% criterion rate.

Software applications

Although the processing of a Monte–Carlo analysis is fairly straightforward it does involve serious number crunching and the only viable means is to use a customised or proprietary software package. There are numerous alternatives available.

Taking an example INFRISK has been developed by an in-house team within the Economic Development Institute of the World Bank. It is intended to create greater awareness and expertise concerning the evaluation and management of risk. INFRISK analyses infrastructure projects' exposure to different market, credit and performance risk from the perspective of the different project actors.

From an equity holders viewpoint IRR and NPV are the main matrices. The IRR is generated from user changes and other forms of income, therefore it is assumed that

$$\text{IRR} = f(m, r, l, s, pi)$$

175

where

m = debt maturity

r = interest rate

l = project's debt equity ratio

s = government support, e.g. tax incentives, depreciation allowances, guarantees

pi = tariff charge

From a creditors viewpoint they will be concerned with the project's capacity to attract loans and government will be concerned with social welfare and national benefit.

INFRISK claims to be capable dealing with multiple sources of uncertainty and risk that are applicable to project viability, e.g. revenue streams, operations and maintenance costs and construction costs. The user has the option of selecting applicable risk variables. The variables are then processed using Monte carlo simulation to produce single 'best guess' estimates over the life of the project. This information can then be used to construct probability solutions for each variable. Four classes of probability distribution are used (uniform, normal, log normal and beta) to provide flexibility according to suitability for specific variables. INFRISK recognises that the risks associated with construction will be different to those associated with post-occupation operation and therefore these are dealt with separately.

INFRISK works in conjunction with MS Excel and a user guide is available online at www.worldbank.org/wbi/infrafin/infrisk.html

Financial appraisal of bidders

The process associated with the selection of bidders is covered in Chapter 6; however it is important that the evaluation of the bids is transparent and that the financial aspects of bids are properly considered against other criteria. The expectation is that all bids submitted will be credible and worthy of serious consideration taking into consideration the need to commit resources, expertise and commitment to the project.

Where bids involve design and build, finance and operation of the project over a concessionary period then there will be a range of financial issues to be taken into consideration. These will include the availability of adequate equity and firm commitments from banks and other lending institutions to provide the necessary loan finance. The strength of these commitments will be judged by the debt:equity ratio, which ideally should fall within the range of 70:30 to 90:10. Essentially the higher the equity the greater the financial stability of the project because of the protection provided to the debt liability associated with the projected stream of revenue from the project. Examples in the previous section of this chapter express the importance of the ability of the project bid to service borrowing and cover all costs leaving an adequate surplus to provide a ROI. This is vital to the interests of all concerned, especially in the case of DBFO projects and other concessionary arrangements spanning long periods of up to 30 years or more.

The financial appraisal of project bids is of prime importance, but there will be other criteria that may be difficult to quantify because of their intangible nature. The difficulty in establishing a common basis for assessment should not be underestimated. This is normally achieved by keeping the criteria to an acceptable minimum and the ability to articulate meaningful relative weightings to each of the criteria. The method of allocating the weighting scores is also important.

The criteria selected should have a direct relationship with the objectives of the project and should be capable of being measured to provide meaningful comparison between alternative proposals. The criteria and weightings for a typical road tunnel project may be as shown in Figure 7.4.

Provision of a new cross mountain road tunnel				
Major objectives	Weighting W1	Decision Criteria	Weighting W2	Combined Weighting W1W2
Financial stability	0.45	User income	0.4	0.180
		Period of concession	0.1	0.045
		Corp. financian str. & stab.	0.2	0.090
		Op. & maint. Costs	0.3	0.135
Design & construction quality	0.20	Design excellence	0.6	0.120
		Constr. Methods & del syst.	0.4	0.080
In use performance	0.20	Sust. toll collection	0.5	0.100
		Op. & maint. Proposals	0.5	0.100
Delivery timescale	0.15	Construction period	0.7	0.105
		Timing of phases	0.3	0.045
			Total	1.000

Figure 7.4   Project appraisal criteria and weightings.

Having established the relative weighting of the criteria the next step is to score each proposal against the assessment criteria using a score of 0–100, where 0 represents no contribution and 100 represents the maximum contribution. The comparison of three project proposals is shown in Figure 7.5.

Assessment Criteria	Weighting (W)	Proposal 1		Proposal 2		Proposal 3	
		Score (S)	WS	Score (S)	WS	Score (S)	WS
User income	0.180	65	11.700	40	7.200	80	14.400
Period of concession	0.045	40	1.800	30	1.350	70	3.150
Core finance - str. & stab.	0.090	25	2.250	70	6.300	60	5.400
Op. & maintenance costs	0.135	60	8.100	65	8.775	65	8.775
Design excellence	0.120	70	8.400	55	6.600	70	8.400
Construction methods	0.080	90	7.200	70	5.600	60	4.800
Sust. toll collection	0.100	65	6.500	70	7.000	75	7.500
Maintenance proposals	0.100	45	4.500	65	6.500	80	8.000
Construction period	0.105	80	8.400	50	5.250	90	9.450
Timing of phases	0.045	85	3.825	40	1.800	85	3.825
Total			62.675		56.375		73.700

Figure 7.5   Project comparison.

From the afore analysis Proposal 3 is clearly the strongest. However, it is argued that this approach relies heavily on subjective assessment in establishing the weighting of the criteria and the allocation of scores. Nevertheless the analysis provides an indication to assist final judgement on the options presented. This technique also assumes that each of the criteria is mutually independent, but in reality this may not be the case.

There are other more sophisticated techniques involving pairwise comparisons using analytical hierarchy process (AHP) (Saaty, 1980) or step-wise procedure (Kepner, 1994) and fuzzy logic (McNeil, 1994).

Project financial monitoring and control

Adequate project financial control is an important pre-requisite of a successful and profitable infrastructure project. At the design stage strict control should be maintained over expenditure by the implementation of robust monitoring designed to provide early warning of adverse cost trends. Provision should be made for timely reporting to the management decision-making system to facilitate action that will eliminate or mitigate problems or difficulties that have the potential to cause additional unnecessary expenditure. The monitoring and control systems required for decisions associated with design will need two dimensions, namely controlling the design budget and predicted life cycle costs, and secondly, the control of expenditure associate directly with the resources necessary to create and fully document the design to facilitate construction. The control of construction costs will be primarily associated with the efficient deployment and control of resources and the management of the construction supply chain to achieve optimal productivity.

In the event that the project involves a concessionary arrangement spanning over a period of years, where, for example, a PPP contractor is required to operate and manage the project, then competent systems will be necessary to control operating costs. Hence, principles associated with good facility management will come into play and steps should be taken that will minimise necessary expenditure required to achieve agreed benchmarks of performance in use.

The three distinct stages normally associated with an infrastructure project are design, construction and postoccupation/handover. Each stage will have different financial control requirements and these need to be accommodated by the generation of specific cost information and data that should be capable of being integrated to control the project as a whole.

Monitoring and control of the design

At the inception of the project early estimates will be needed to establish whether the requirements of the client are feasible. For this purpose it is usual to utilise cost data collected from past projects alongside data published concerning similar projects previously constructed across the industry. Accepted cost planning practice is used to adjust such data to suit the project in hand. As the design progresses from outline to a more detailed scheme, the project will be more accurately costed in accordance with its elemental breakdown. It is usual to express such costs as items, linear, superficial or cubic measurements. Alternative designs will be sought and compared to achieve an optimal solution, given the information available.

As a project evolves and takes shape, costing will be extended to include the operation of the project from occupation to the completion of its design life. There is an increasing awareness amongst clients that lifecycle costs far exceed costs associated with design and construction which normally only represent approximately 10% of total expenditure. It is therefore not surprising that clients are becoming more sophisticated in their consideration of the operating costs predicted for specific designs. The growing popularity of procuring projects using PPP has also engaged the interest of concessionary contractors operating BOT type contracts where they are required to take on risk associated with the performance of the design in use. The underlying issues concerned with effective optimisation of infrastructure design according to client requirements and costs in use are as follows (Alexander, 1996):

- fulfilment of the client brief and expectations;
- innovation in design and the application of appropriate and up to date technology;
- sufficient past cost and performance data and corporate knowledge from in-house and external sources;
- procedures for measuring the performance of the project regarding energy consumption, sustainability and conformance with accepted benchmarks;
- adequate procedures for evaluating risk and associated techniques for comparative analysis, especially that required for the appraisal of life cycle costs;
- monitoring the design against the design budget as it evolves; and
- allocation and control of resources allocated to the design team.

Monitoring and control of the construction process

Whatever procurement process is used the organisation responsible for the construction process will need to prepare a method statement and a project master programme. These will form a basis for the determination of the construction budget. In order to exert close financial control it will be necessary to identify a work breakdown structure that can be grouped into work packages or cost centres. The work packages will be further analysed and broken down, as appropriate, into resource types and the level of breakdown can be more detailed by identifying discrete activities, locations and working areas. In this manner expenditure can be allocated and compared with the value of work completed according to the budget.

Adequate project control relics on a competent project control system that provides for the monitoring of time and cost, together with resource utilisation if required. Ideally this facility needs to be integrated with the project's accounting system to ensure that double handling and duplication of data is avoided. Allowance must be made for additions and omissions to the original budget before an assessment can be made of profitability and progress to date. The consequences of not achieving sufficient integration will be likely to result in a cumbersome, less reliable and more expensive control system, which in extreme cases may cost more than the potential savings to be achieved. It is therefore essential that adequate consideration is given to the selection and prior evaluation of a suitable cost control system and its associated software that meets both project and corporate requirements.

Earned value analysis (EVA) is a technique that provides a complete evaluation of a project as it progresses and predictions can be made regarding the final project cost and the period required to complete the project. The original budget known as 'budgeted cost of work scheduled' (BCWS) which normally takes the form of an 's' curve. As the work progresses the 'budgeted cost of work performed' (BCWP) can be established at an early stage. The difference between BCWS and BCWP represents the variation in the schedule (SV) and the ratio created by BCWP/BCWS provides the schedule performance index (SPI). In the event of a negative SV and an SPI of less than unity a time overrun is indicated and the project duration divided by the SPI provides a revised estimate for the project to be completed (Figure 7.6).

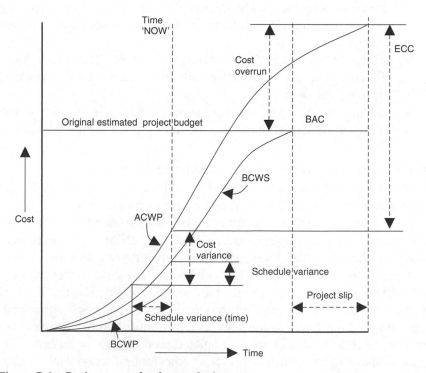

Figure 7.6   Project earned value analysis.

The actual cost of work performed (ACWP) is determined from data provided by the project cost control system. The cost variance (CV) is represented by the difference between BCWP and ACWP and the ratio generated by BCWP/ACWP gives the cost performance index (CPI). A negative CV and a CPI of less than unity indicates a cost overrun, conversely a positive values show that the project cost performance is better than expected.

The estimated cost to complete the project (ECC) when added to the ACWP provides the final cost to completion. This method represents a straight forward approach that provides consistency to enable comparison between different projects by using the project indices CV and CPI. Care needs to be taken when comparing SV and SPI because project activity sequences are not taken into account (Howes, 2000).

Monitoring and control of operational costs post occupation

An important prerequisite of effective control of operational costs is the preparation of a comprehensive budget that allows sufficient resources to permit the achievement of agreed project performance benchmarks (Spedding, 1994). Similarly budgets should be prepared for planned maintenance and repairs. Where a project is procured under a PPP/BOT type of arrangement then detailed budgets and risk assessments will have normally already been undertaken at the design stage, and in the case of a facility management contract budgets should have already have been prepared as part of the bid. It is important that the cost control system integrates with the database provided by the accounting and project control systems.

Typically budgets will have been prepared under the following headings:

- energy consumption;
- external services – water, sewage, waste disposal;
- administration and management;
- cleaning;
- replacement;
- annual and periodic maintenance;
- security;
- taxation; and
- insurance.

Subheadings can be employed according to the degree of control required and the nature of the project. Procedures should be established for the efficient collection and input of cost data in a form that can readily be used. The cost control system must be capable of providing holistic and appropriately detailed reports according to cost headings, resources and

181

companies in the supply chain and by project location. Ideally the financial control system should have facilities to predict final expenditure, as well as being capable of undertaking various forms of analysis, e.g. sensitivity, risk and 'what if'. Archiving facilities are also desirable for the purpose of incrementing the corporate knowledge base according to the experience gained by the project. The design of control reports should enable timely decision-making derived from reliable and relevant cost and performance data.

Case studies

Case study 7.1: Ohio Department of Transport (ODOT), USA

The state of Ohio has the 10th largest road network in the USA and it has the 5th largest volume of traffic. The state depends heavily on taxation generated from the sale of gasoline to fund its escalating system of maintenance requirements. However, road users are taking advantage of more environmentally friendly ethanol-based fuels, which have a lower tax levy and consumption has risen from 16 to 40% of total gasoline consumption in the past 5 years. This is having a constraining effect on the state's ability to fund new projects, since 85% of ODOT's revenue is taken up by preserving the large existing state road network. To help overcome this problem ODOT has introduced a range of new leveraging techniques.

This case study focuses on ODOT's use of toll credits to help match the Federal share provided by Grant Application Revenue Vehicles (GARVEE) for accelerated construction from bond issues.

Any highway or transportation project is eligible under the State Transportation Improvement Programme and must comply with the National Environmental Protection Act. GARVEE bonds have a 30 year maximum term and normally funds are dispersed on a reimbursement basis and must be matched from other identified funding sources. It is possible for a discriminatory initial deposit to be made under certain circumstances.

The use of GARVEE bonds in Ohio in combination with toll credits, has enabled the advance of three major transportation projects, namely the Spring-Sanduskey Corridor, the Maunee River Crossing and the South east Ohio Plan. By combining these two innovative financing tools ODOT has optimised limited transportation dollars and increased investment in projects vital to a free flowing state road network. Major projects have been expedited years ahead of what could have been achieved using traditional financing techniques. In addition, toll credits have been freed to provide cash resources for allocation to system maintenance requirements and other priorities.

Author's commentary

By the use of innovative funding techniques that have been properly thought out, ODOT has been able to bring forward major infrastructure projects far in advance than would have otherwise been the case using traditional financing methods. This case has demonstrated the importance of being able to raise match funding and to justify the worth of such projects regarding their economic, social and environmental impact and benefits.

Case study 7.2: The Rion Antirion Bridge, Greece

The Rion Antirion Bridge is a 2.52 km, 4 pylon cable-stayed bridge with a span distribution of 286, 560,560 and 286 m respectively. There are two approach viaducts, 392 m on the Rion side and 239 m on the Antirion side. The deck is 27.2 m wide and is fully suspended throughout its total length. The bridge has been designed to withstand seismic shocks and impacts from large ships.

The total cost of the project amounts to €800 million and is judged to be of high economic and social importance to the region. It has therefore qualified for a loan from the European Investment Bank (EIB), guaranteed by a pool of commercial banks, amounting to 45% of the total cost. Remaining funding is provided by the Greek State at 45%, together with 10% from share capital.

The EIB master facility agreement is between Gefyra S.A. and the EIB, which has granted to the Concessionaire a €370 million floating interest rate loan. The loan can be drawn down during the construction period according to the funding needs of the concessionaire and is to be repaid in full no later than 25 years from the last disbursement. The first draw down is dependent on the Checker's final approval of the design of the bridge.

Gefyra SA is a Greek JV Company established in 1995 by VINCI (France) and six other Greek contractors with the sole purpose of entering into a concession contract for the construction of the Rion Antirion Bridge. The breakdown of ownership is as follows:

VINCI	53.00%
Elliniki Technodomiki–Telo SA	15.48%
J & P – AVAX SA	11.20%
Athena SA	7.74%
Proodeftiki SA	7.74%
Pantechniki SA	4.84%

The parties to the project are as follows:

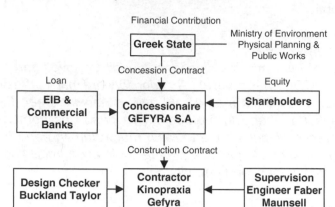

Geyfra SA is responsible for the design, construction, financing, maintenance and operation of the bridge during the 42 year concession period. Compliance with specified commitments has required numerous agreements, including the design and build contract with the contractor and extensive financing documentation with the creditors.

Author's commentary

The Rion Antirion Bridge is an excellent example of a large infrastructure project being undertaken by an international joint venture with the express intention of becoming the concessionaire for this project under a DBFO arrangement. The project demonstrates the importance of the EIB loan agreement and its contribution to the financial feasibility of the project.

Summary

This chapter illustrates the importance of adequate and appropriate financing of infrastructure projects, which is an essential prerequisite to successful project procurement and implementation in support of predetermined aims and objectives. The strategy and policy adopted by government sets the agenda for PPP and the potential to leverage private investment into public projects. The financial demands placed on concessionaires have been explored and the need to undertake adequate project analysis to evaluate feasibility and risk has been highlighted. Given that concessionary periods span over typically up to 30 years various techniques associated with DCF and NPV and uncertainty have been explored to assist investment decision-making and to help mitigate risk.

The financial appraisal of bidders has been investigated and the importance of selecting the winning bid based on sound criteria other than the cheapest price has been described. The chapter concludes with an appraisal of project cost monitoring and control at all stages of the project and an overview is provided of the necessary actions to implement an effective financial and cost control system.

The next chapter specifically addresses the issues and procedures associated with financing projects in developing countries.

References

Akintoye, A., Beck, M. and Hardcastle, C. (2003), '*Public–private partnerships*', Blackwell Science, Oxford, UK.

Alexander, K. (1996), '*Facilities management: theory and practice*', E & FN Spon, London (ISBN 0-419-20580-2).

Construction Industry Council (1998), '*Constructors' guide to PFI*', Thomas Telford, London.

Damoderan, A. (2001), '*Corporate finance theory and practice*', 2nd Ed., Wiley International, New York.

Farrall, S. (1999), '*Financing transport infrastructure: policies and practice in Western Europe*', MacMillan, London.

Howes, R. (2000), '*Improving the performance of earned value analysis as a construction project management tool*', Engineering, Construction and Architectural Management, ISSN 0969 9988, 7(4), 399–411, Blackwell Science, Oxford UK.

Kepner, C.H. (1994), '*The rational manager: a systematic approach to problem solving and decision making*', 2nd Ed., Kepner Tregoe, Princeton NJ.

Levy, S.M. (1996), '*BOT: paving the way for tomorrow's infrastructure*', Wiley, New York.

McNeil, F.M. (1994), '*Fuzzy logic: a practical approach*', AP Professional, UK.

Merna, T. and Njuru, C. (2002), '*Financing infrastructure projects*', Thomas Telford, London.

Miller, J.B. (2000), '*Principles and practice of public and private infrastructure delivery*', Kluwer Academic Publications, Dordrecht.

'*PFI BOT Promotion*', Project Finance International, No. 93, March 1996.

Saaty, T.L. (1980), '*The analytical hierarchy process: planning, priority setting, resource allocation*', McGraw Hill International Book Co., New York.

Samuels, J.M. and Wilkes, F.M. (1986), '*Management of company finance*', 4th Ed., Van Nostrand Reinhold, UK.

Smith, A.J. (1999), '*Privatised infrastructure: the role of government*', Thomas Telford, London.

Taha, H.A. (1982), '*Operations research: an introduction*', 3rd Ed., MacMillan Publishing, New York.

Walker, C. and Smith, A.J. (1995), '*Privatised infrastructure; the build operate and transfer approach*', Thomas Telford, London.

Walters, C.D.J. (1989), '*A practical introduction to management science*', Addison-Wesley Publishing Co., Reading, MA.

Chapter 8

Funding infrastructure projects in developing countries

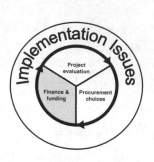

Introduction

A key factor influencing the economic growth of developing countries is the amount and quality of infrastructure provided for transport, water, power, sewage, waste disposal, education and health. Such provision is costly and normally requires expertise and resources that are often not available locally. This necessitates a long-term strategic view, whereby projects are prioritized and brought forward as and when they can be afforded. Unfortunately, some less well off developing countries often suffer disruption caused by internal conflicts, corruption and a lack of political continuity that makes long-term planning extremely difficult. As a consequence, these countries fall further behind their counterparts who have more stable socio-economic and political circumstances. Hence, resources available for the improvement and development of infrastructure are often beyond the financial capacity of their respective governments.

Most developed countries have bilateral aid agencies, specialist agencies and trust funds that provide financial support to developing and least developed countries to alleviate poverty and to improve living standards as well as for export promotion. In addition, governments from the most developed countries contribute funds to multilateral development and donor agencies who then allocate funds to the governments of developing and least developed countries. Without these development agencies, least developed countries would be severely affected and potentially condemned to extreme poverty and hardship.

This chapter discusses the need and mechanisms for funding infrastructure development in developing countries. The structure and composition of the major multilateral and bilateral funding agencies is explained, together with the means by which they allocate and distribute funds for infrastructure development. The role of the World Bank and the regional development banks and their contributions to infrastructure development are explained with specific reference to the need for leverage in private investment. The concept of co-financing as an alternative way of funding infrastructure projects is discussed. The need to identify, manage, reduce or eliminate conflicting issues in co-financing processes is highlighted as essential in improving the viability of infrastructure projects. Attention is also given to the need to have adequate project identification capacity to screen projects for funding and for a thorough understanding of the impact of risks in evaluating funding for infrastructure projects in developing countries.

The need for funding infrastructure

Developing countries have huge requirements for infrastructure development in order to support growth, reduce poverty and improve living standards. There is an urgent need to increase funding for capital and recurrent investment to improve, or at worst, to slow down the deterioration of existing infrastructure.

Access to infrastructure services is one of the important factors in defining welfare as it affects the dimensions of poverty in terms of employment and income (Mabogunje, 1993; Daniere, 1996; Sethuraman, 1997). The low level of social development, exacerbated by rapid population and urbanisation growth rates in developing countries has, and will continue to create poverty and generate enormous demand for infrastructure. There is, therefore, a need for substantial funding to address poverty issues and infrastructure deficits. Figure 8.1 shows the extent of poverty in the developing world.

Region	1985	1990	2000
All developing countries	30.5	29.7	24.1
South Asia	51.8	49.0	36.9
East Asia	13.2	11.3	4.2
Sub-Saharan Africa	47.6	47.8	49.7
Middle East and North Africa	30.6	33.1	30.6
Eastern Europe [a]	7.1	7.1	5.8
Latin America and the Caribbean	22.4	25.5	24.9

Figure 8.1   Percentage of population below the poverty line in the developing world, 1985–2000.
Source: World Bank.
[a] Does not include the former U.S.S.R.

The question of how to increase investment to expand infrastructure provision to provide vital services, particularly for industrial productivity, has now become a major challenge and a central policy issue. According to the International Finance Corporation (IFC), developing countries will require more than US$ 3 trillion for investment in new infrastructure over the next 10 years. Estimates for developing the Asian region range from US$ 1 to US$ 2 trillion, US$ 600 billion for Latin America while Eastern Europe and Africa also need heavy capital infusions for infrastructure development. The situation is made worse for recurrent financing, particularly in projects requiring foreign exchange, as the involvement of funding agencies does not normally extend to the operation and maintenance stages. This point is confirmed by (Gaude and Watzlawick, 1992) who noted that the contributions of donors and development banks are often restricted to inputs to design and construction works, while expecting national sources to share costs for operation and maintenance. The need for the involvement of donors in the operation and maintenance stages is implicit in World Bank's argument that lending should be conditional on an acceptable distribution of recurrent cost expenditure for maintaining existing infrastructure and to building new projects (World Bank, 1988). Infrastructure provision will, therefore, remain low in developing countries, and further investments will be lost unless fundamental issues concerning funding are addressed.

Bilateral and multilateral funding

Where governments lack sufficient resources to fund infrastructure projects they will either cancel or delay projects, or alternatively will turn to an external source to acquire loans for the purpose of making up the shortfall. In the first instance indigenous private sources will be sought after, but if these prove to be inadequate then the assistance of aid agencies will normally be investigated.

Development agencies are classified as multilateral where governments contribute to an international organization such as the World Bank, or bilateral where a single country has a specific programme to assist less well off countries. Figure 8.2 provides a list of the main agencies and their geographic area of coverage. The structure of the leading multilateral development agencies – World Bank, United Nations Development Programme, African Development Bank and Asian Development Bank are outlined in the next sections.

MULTILATERAL		BILATERAL		
Agency	Coverage	Country	Agency	Coverage
World Bank IAD/IBRD	Global	Canada	CIDA	Global
U.N. Development Program	Global	Denmark	DANIDA	SSA
Asian Development Bank (AsDB)	Asia	France	AFD	N & W Africa
Inter American Dev. Bank (IADB)	Central & S. America	Germany	GTZ	Mainly Africa
African Development Bank (AfDB)	Africa	Japan	JBIC	Mainly Asia
European Bank for Reconstruction (EBRD)	Eastern Europe & Russia	Norway	NORAD	Mainly Africa
EU (PHARE & TACIS)	Ditto	Sweden	SIDA	SSA & S. Asia
Caribbean Development Bank (CDB)	Carribbean Region	Switzerland	SDC	Global
		UK	DFID	Commonwealth
		USA	USAID	Global

Figure 8.2   Multilateral and bilateral agencies.
Source: World Bank (2002).

World Bank

In 1944, the World Bank was formed as the International Bank for Reconstruction and Development (IBRD). Today the IBRD concentrates on providing loans and development assistance to middle income countries and has been joined by the International Development Association (IDA) that provides concessionary loans to very poor countries defined as having a GDP per capita of less than US$ 885 (World Bank, 2002). World Bank also consists of the International Finance Corporation (IFC), the Multilateral Investment Guarantee Agency (MIGA) and the International Centre for Settlement of Investment Disputes (ICSID) (see Figure 8.3).

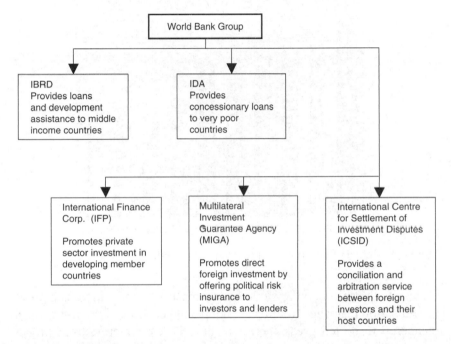

Figure 8.3   World Bank Organisation.

To qualify for IDA funding it is necessary for applicants to meet the following criteria:

1. Eligibility – applicants must be able to demonstrate relative poverty as defined by GNP per capita criteria and a lack of credit worthiness to borrow on market terms and good policy performance.
2. Performance rating – on an annual basis World Bank staff assess the quality of each borrowers performance.
3. Lending and performance – IDA management staff monitor lending to each country and actual lending per capita is correlated with performance levels.

Infrastructure investment lending by the World Bank has fallen significantly between 1993 and 2002. The combined spending of IDA and IBRD

189

stood at US$ 9.5 billion in 1990, whereas this had declined to just over US$ 5 billion in 2002. Figure 8.4 illustrates how the IDA's share of spending has actually increased, while IBRD loans have dramatically reduced. The reason for this is that the World Bank's infrastructure business policy has shifted from 'bricks and mortar' to a model of service delivery, where sustainable issues are taken into account. Hence, in more well off developing countries there has been growth in private investment in infrastructure, thereby reducing the need for IBRD loans.

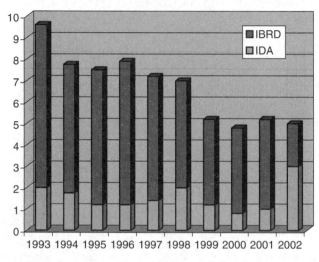

Figure 8.4   IDA/IBRD infrastructure investment lending (US$ billion).

Despite the shift in policy to leverage in private funds, there is a growing realisation that where returns are not perceived to be high, or the risk is considered to be too great, there will be a lack of interest from private investors. This is particularly the case in poor countries that have less than stable financial and political climates. Hence, the World Bank's overarching policy is to leverage funds from an entire spectrum of public and private sources supported by IBRD, IDA, IFC and MIGA products. It is therefore important that project proposals requesting loans are adequately prepared and justified.

In countries that qualify for IDA loans, private sector investment will continue to be encouraged, however, it must be recognised that this will remain low for the foreseeable future. Consequently, the IDA will remain an important source of borrowing alongside regional development banks and other multilateral and bilateral agencies.

United Nations Development Program (UNDP)

The UNDP operates in accordance with the principles and values of the United Nations. Its mission is to provide advice and support for developing countries covering an entire range of issues associated with poverty reduction, institutional capacity building and coping with the challenges of globalisation. The UNDP is the United Nation's main supplier of development advice and grant support. In 2003, its core funding amounted to US$ 36 million, however, by adopting a partial funding system formula, the total value of project funding far exceeded this figure. The UNDP's core programmes focus on the countries that have 90% of the world's extremely poor people.

The UNDP is controlled by its Executive Board (EB), which includes representatives from 36 nations around the world on a rotating basis. The EB is responsible for ensuring the relevance and implementation of its policies and for providing inter-governmental support. The UNDP has in the region of 130 field offices, which represents a presence in nearly every developing country.

African Development Bank (AfDB)

The AfDB established in 1964, is a multilateral development bank that represents the economic development and social interests of its 53 regional member countries supported by 24 stakeholder countries in North and South America, Asia and Europe. The prime roles of the bank are to:

- provide loans and equity investments for the economic and social enhancement of its regional members;
- provide technical assistance and advice for the initiation and implementation of development projects;
- promote investment in public and private capital for the purpose of development;
- respond to requests for advice and assistance to coordinate regional development policies and plans; and
- encourage and support regional integration between member countries.

Figure 8.5 shows the total disbursement of loans from all funds between 2001 and 2002. The allocation of loans to infrastructure has increased over this period; but the percentage allocation has fallen by 10% to 39.54%. Major increases have been made in the finance and agricultural sectors.

The bank lends at a variable lending rate calculated on the cost of its borrowings. The rate is adjusted twice per year on 1st January and 1st July, respectively. Maturities range up to 20 years with a 5-year grace period and a 1% commitment charge.

The African Development Fund (ADF) provides loans on concessionary terms to low-income members. The Fund finances projects, specific studies and technical assistance. It lends at no interest rate, with a 0.75% per annum service charge, a 0.5% commitment charge and a 50-year repayment period, including a 10-year grace period. The Nigeria Trust Fund (NTF) provides

191

Loans and Grant Disbursements by Sector (US$ Million)			
Sector	2000	2001	2002
Agriculture	160.45	163.51	200.65
Transport	172.07	196.83	170.73
Communications	60.04	2.54	51.73
Water & Sanitation	69.63	53.96	71.25
Power	56.86	71.50	61.05
Industry, mining & quarrying	18.54	38.30	21.23
Finance	136.82	202.82	272.30
Education	64.40	47.42	79.75
Health	35.74	34.58	64.98
Poverty Alleviation	7.68	30.60	25.64
Gender, pop. & nutrition	2.50	1.37	0.97
Other social sector	0.00	0.07	0.50
Urban development	0.00	0.00	0.00
Environment	0.41	0.42	0.78
Multisector	177.71	326.29	403.80
Total Infrastructure	**476.54**	**429.38**	**563.60**
Infrastr. % of Total	**49.49**	**36.69**	**39.54**
Total	**962.85**	**1170.23**	**1425.36**

Figure 8.5   Loan disbursement by the African Development Bank.
Note: Multisector allocated according % infrastr. spending of total.
Source: African Development Bank (2003).

financing for projects of national and regional importance in the furtherance of economic and social development of low-income members.

The prime objective of the AfDB's Strategic Plan 2003–2007 is to achieve better development effectiveness and to improve the benefit derived through loans by better monitoring and control procedures. The bank intends to give priority to agriculture and sustainable rural development, physical infrastructure with greater emphasis on economic infrastructure such as energy, water supply and transport, and human capital formation (personal infrastructure) through primary education and basic health services. Selective support will also be given to the development of essential infrastructure for urban and rural development. Special premium will also be given to private sector developments and capacity building initiatives and programmes that bring the benefits of globalisation to members, especially through direct foreign investment and the promotion of SMEs.

Asian Development Bank (AsDB)

The AsDB is a non-profit making multilateral financial institution and it is owned by 62 members, mostly from the region. Its headquarters is in Manilla and it has 24 other offices located primarily in Asia, together with representative offices in Frankfurt for Europe, Tokyo for Japan and Washington DC for North America.

The AsDB's prime objective is to reduce poverty in Asia and the Pacific by promoting pro-poor sustainable economic growth, social development and good governance. In support of these ideals the AsDB concentrates on:

- protection of the environment;
- protection of gender and development;
- private sector development; and
- regional cooperation.

The AsDB raises funds by issuing bonds on the world's capital markets, as well as contributions from members.

Funding arrangements are divided into categories according to the wealth of borrowing countries. An ordinary capital resource (OCR) is a pool fund that makes loans at near market terms to better-off borrowing countries. The Asian Development Fund (ADF) provides 'concessional' or 'soft' loans using funding provided by donor member countries. ADF loans have very low interest rates and are intended for the poorest borrowing countries. A special fund is also available specifically for poverty reduction (JFPR) and as of October 2002 the total amount of approved projects was US$ 68.73 million.

Figure 8.6 contains the following tables that provide details of the loans made by lending, Loans by sector – 2002, and Borrowers by Country.

ADB Loan Approvals (US$ Millions)	
Year	Amount
1998	5,982
1999	4,979
2000	5,583
2001	5,339
2002	5,676

ADB Loans by Sector – 2002		
Sector	US$ Millions	%
Transport & Communications	1612.90	28.4
Energy	1017.60	17.9
Finance	865.00	15.2
Social Infrastructure	669.81	11.8
Agriculture & Natural Resources	492.90	8.7
Multisector	155.32	2.7
Industry & Nonfuel Minerals	85.00	1.5
Others	777.22	13.7
Total	5675.75	100.0

Figure 8.6   Tables indicating the amount and nature of AsDB borrowing.
Source: Asia Development Bank (2003).

In order to build expertise and capacity, the AsDB provides grants typified by the following tables shown in Figure 8.7.

AsDB Technical Assistance US$ Millions	
Year	Amount
1998	148
1999	171
2000	170
2001	148
2002	178

Technical Assistance Grant by Sector – 2002		
Sector	US$ Millions	%
Social Infrastructure	22.06	12.3
Multisector	20.25	11.3
Finance	17.48	9.8
Transport & Commun.	16.24	9.1
Agri. & Natural Resources	15.79	8.8
Energy	11.49	6.4
Ind. & Nonfuel Minerals	6.33	3.4
Others	28.82	16.1
Total Grants to DMCs	138.46	77.2
Regional Activities	40.58	22.8
Total	**179.04**	**100.0**

Recipients of Technical Assistance Grants – 2002		
Country	US$ Millions	%
Indonesia	19.10	10.7
Afghanistan	15.14	8.5
India	13.24	7.4
PRC	13.20	7.4
Vietnam	9.28	5.2
Pakistan	7.67	4.3
Phillippines	6.60	3.7
Cambodia	6.53	3.6
Bangladesh	4.85	2.7
Sri Lanka	4.79	2.7
Other DMCs	38.08	21.3
Regional Activities	40.58	22.7
Total	**116.50**	**100.0**

Figure 8.7 AsDB Technical Assistance.
Source: Asia Development Bank (2003). *Includes loan components of regional projects to Cambodia, Kyrgyz Republic, Lao People's Democratic Republic, Tajikistan, Uzbekistan, and Vietnam.

Borrowers by Country – 2002		
Country	US$ Millions	%
India	1183.60	20.9
Pakistan	1141.00	20.1
PRC	868.48	15.3
Indonesia	767.22	13.5
Viet Nam	315.00	5.5
Bangladesh	299.77	5.3
Sri Lanka	236.50	4.2
Uzbekistan	166.50	2.9
Cambodia	100.91	1.8
Other Developing	296.77	5.2
Member Countries	150.00	2.6
Regional Activities*		
Grand Total	5675.75	100.0

Figure 8.7   *Continued*

There are three instruments through which grants and loans for technical assistance are made:

- *Project Preparatory Technical Assistance (PPTA)*: intended to assist developing member countries to establish a series of investment projects suitable for AsDB or other funding. This comprises primarily feasibility studies and may also include capacity building, promotion of new financial markets and policy reform.
- *Advisory and Operational Technical Assistance (AOTA)*: to assist DMCs to undertake proposal evaluation, contract supervision and the management of AsDB-financed projects. This activity will also include capacity building and the development of national and sectoral policies.
- *Regional Technical Assistance (RETA)*: intended to promote regional activities, including studies, conferences, workshops, research and other relevant activities.

Good governance is a major part of the AsDB's strategy that may challenge established rules of a constitutional nature. The AsDB promotes sound development through transparency, accountability, efficiency and participation.

Loan applications

Applications for project loans to be considered by investment banks usually progress through a sequence of phases that eventually lead to rejection or acceptance and the implementation of a project. These will vary according to practice and procedure, however, the World Bank project cycle shown in Figure 8.8 represents the outline of a typical example.

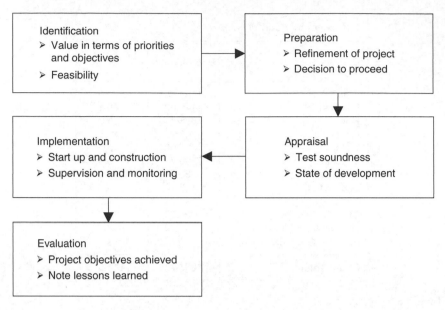

Figure 8.8   Project cycle.

Private sector funding

There is a growing expectation among multilateral and bilateral aid agencies that the private sector should be encouraged to invest in large infrastructure projects. For example, the International Finance Corporation (IFC) seeks to ensure participation from the private sector by limiting the amount of own-account debt and equity financing it will provide for any single project. In the case of new projects, a maximum of 25% of the total project cost is allowed with up to 35% for smaller projects. In the case of expansion projects, the IFC will provide up to 50% of the total project costs on the assumption that investments do not exceed 25% of the capitalization of the project company.

The major problem is that private investors will be aware of the broad spectrum of opportunities, each of which will be assessed regarding the potential to provide a return that is set against the risks involved. The case for investing in infrastructure projects is further complicated by the amount of equity required and the period necessary to provide an acceptable return on investment. Where co-financing arrangements are in place then there might be conflicts of interest among investors, which if apparent from the outset will serve to dissuade potential private investors. The private sector will also be looking for a stable socio-economic and political environment with good future prospects for economic growth and development.

Private finance is normally derived internally from domestic sources, or externally from foreign sources. Domestic finance is usually provided through national development banks, development agencies such as local NGOs and other private sources. External finance can be secured from multilateral development agencies, such as the IFC, international commercial banks as well as other private sources from abroad (see Table 8.1).

Table 8.1   Financing sources for private sector infrastructure

Domestic sources	External sources
Equity	
Domestic developers (independently or in collaboration with international developers)	International developers (independently or in collaboration with domestic developers)
Public utilities (taking minority holdings)	Equipment suppliers (in collaboration with domestic or international developers)
Other institutional investors (likely to be very limited)	Dedicated infrastructure funds
	Other international equity investors
	Multilateral agencies (International Finance corporation, Asian Development Bank)
Debt	
Domestic commercial banks (3–5 years)	International commercial banks (7–10 years)
Domestic term lending institutions (7–10 years)	Export credit agencies (7–10 years)
Domestic bond markets (7–10 years)	International bond markets (10–30 years)
Specialised infrastructure financing institutions	Multilateral agencies (15–20 years)
	Bilateral aid agencies

Source: Ahluwalia (1997, p. 97).

Infrastructure projects usually require huge investments that are often outside the capacity of the domestic private sector in developing countries. Financing is crucial for infrastructure development, but only a relatively small proportion of infrastructure in developing countries is financed locally (IFC, 1996). According to Gaude and Watzlawick (1992), the proportion of externally financed public infrastructure investments in the least well off developing countries generally exceeds 50% and can be as high as 70–90%. A significant proportion of equity financing is required for projects with higher levels of perceived risks. Kisanga (1996) noted that private finance in developing countries is likely to be obtained from foreign sources, in the short and medium term.

External debt financing is likely to be available only to countries with reasonable credit ratings and well-structured private sector projects. Also the proportion of equity to debt financing depends on the sectors. Ahluwalia (1997) noted that telecommunication projects with relatively high market risks require a relatively low debt component (debt to equity ratios close to

1:1), while power projects with assured power purchase agreements would be financed with debt to equity ratios of 2.5:1 or even 3:1.

The potential and need for private investment will increase as institutional reforms are implemented, and as public expenditure and borrowing continues to be tightened due to the conditions imposed by structural adjustment programmes. But the scope for private investment in infrastructure does not only depend on the economic and political commitment, but on investors' interest. Although private finance is becoming increasingly important in infrastructure development in developing economies, opportunities are limited due to a lack of investor interest. Private finance requires a commercial return at acceptable risk. Investors tend to prefer foreign exchange earning infrastructure projects where demand is growing rapidly and there is clear government support and policy environment (IFC, 1996).

The balance between public and private sector investment is also becoming increasingly important. Traditionally, international development agencies have concentrated mainly on financing public projects but major development banks have been reassessing the relative roles of the private and public sector in developing countries. The IFC is actively encouraging new products that involve innovative methods of financing to help boost a general lack of interest from private investors where they perceive returns to be too low and the risks unacceptably high.

Recent trends towards public–private partnerships have resulted in varying degrees of privatisation of infrastructure, which reflects the view that the private sector can make a substantial contribution towards development. Figure 8.9 shows the prime options for private sector involvement

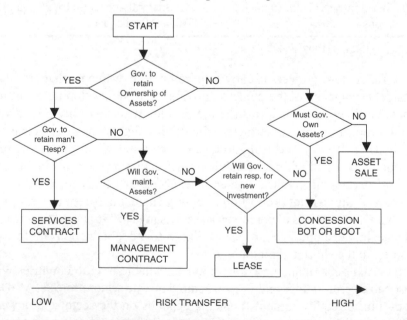

Figure 8.9   Private sector contracts and their associated risk transfer to the private sector.

198

in infrastructure projects and the degree of risk transfer to the private investors. The procurement routes with less risk involvement are relatively well established and form the mainstay of contracting financed through multilateral and bilateral agencies. However, public–private partnerships involving concessionary arrangements are far more demanding regarding income streams and are therefore unlikely to be a feasible proposition for poorer developing countries.

Private sector involvement, through Build, Operate and Transfer (BOT) and variants of it, has contributed to the improvement of infrastructure services in many developing countries. Further, it seems likely that support for private infrastructure finance in developing countries will increase, particularly, for low risk and foreign exchange/revenue generating projects. IFC operations, for example, have enabled private participation in infrastructure, and this is now the fastest growing element in its portfolio accounting for a large proportion of new loan approvals. Private participation in infrastructure has occurred through privatisation of existing public assets and new investment using mainly limited recourse or non-recourse financing schemes, including BOT and Build, Own and Operate (BOO) arrangements.

A more significant factor is that private project holders tend to be more attractive for commercial bank lending as they are often considered to be more commercially aware and productive. This view is partly supported by a recent study, which shows that private owners build infrastructure more efficiently than public managers (IFC, 2003). There is also evidence that infrastructure privatisation may have helped some countries to improve their credibility on international financial markets.

The private project financier is therefore becoming increasingly important due to problems associated with public provision of infrastructure, including delays due to the role government plays in the financing process, complex regulations restricting state and local infrastructure finance, and interference with user charges. However, the record of private lending so far has been unimpressive as both private lenders and private project holders find the legal, political, pricing and accounting framework in developing countries opaque and unpredictable. The choice between private and public will continue to depend on national governments creating an appropriate and acceptable policy framework that provides an investment environment, which is conducive to investment by the private sector.

Co-financing infrastructure

A substantial proportion of infrastructure projects are financed by international lending and by other development agencies, often using a co-financing approach. There are two distinct types of co-financing; 'joint' co-financing and 'parallel' co-financing. 'Joint' co-financing is where several financiers provide loans in some agreed proportions to finance the same set or package of a project, whereas in 'parallel' co-financing, financiers provide loans to finance separate packages of a project. In parallel co-financing each co-financier administers and supervises its own portion of the project based on its procurement rules. Also, in parallel co-financing project size tends to be smaller due to separate packaging, but to interest a larger number of bidders some financiers have provisions for grouping packages into a single lot for procurement purposes only. In joint co-financing, project preparation, evaluation and administration are sometimes delegated to one of the financiers by agreement but the other financiers lending policies and procurement rules are taken into consideration, and in some cases it is a condition for participation in co-financing.

The trend of the 1990s and early 2000s seems to point to an increase of co-financing activities in developing countries (see Table 8.2). This increase in co-financing activity is due to limits on the financial contributions of individual financiers, and investment needs of projects. This is particularly true for large infrastructure projects where financing from one source is often not sufficient. Also, from the funding agencies' standpoint, it is a mechanism for balancing risks, particularly for expensive and high-risk projects. Co-financing from multilateral, bilateral and private sources is therefore widely used for funding infrastructure projects in developing countries.

Table 8.2   BADEA co-funding participants 1975–2001

Details of BADEA's Joint Financing (1975–2001) (US$ millions)

Co-financiers/year	1975–1995	1996	1997	1998	1999	2000	2001	Total
Total Project Costs	9443.30	699.11	530.10	244.85	431.72	339.38	173.75	11862.12
BADEA	1234.40	84.95	64.42	57.50	93.81	64.03	49.34	1648.45
Arab Funds, OPEC/ISDB	1394.10	59.65	88.44	113.93	110.00	141.38	80.03	1987.53
World Bank Group	1376.8[a]	–	85.00	–	–	–	–	1461.80
AsDB/ADF Group	998.92	17.00[b]	47.06	17.00	50.56	51.86	24.23	1206.63[d]
European Union Group	523.41	77.24	73.70	–	50.65	8.96	–	733.96
Western Ind. Countries[c]	1356.90	89.09	129.75	–	15.44	–	–	1591.18
Beneficiary Governments & Local Governments	2558.80	371.17	41.62	56.42	111.26	73.15	20.16	3232.57

[a] Including about US$ 5 million representing the contribution from IFAD.

[b] Including the contribution from the Development Bank of West Africa and the Development Fund for West African Countries.

[c] Including UNDP.

[d] Including the sum of US$ 4.995 million contributed from the Economic Group of West African Countries for the road project linking Ouagadougou-Leo-Ghanaian Borders in the year 2000 and US$ 5.0 million provided by the Development Bank of South Africa for Electricity Generation Project in Mauritius in 2001.

However, there should be a reasonable number of co-financiers, as too many can complicate project development. It is recognised that having too many co-financiers may increase conflicts in infrastructure development, increase interactions among project participants leading to poor communication and overload the project holder's administrative capability, all of which can undermine project performance. For example, separate payment certificates; sponsor-specific progress reports and other types of information during project development may have to be issued separately to each financier – a process that is undoubtedly complicated in joint financing.

Table 8.3 shows that a significant proportion of World Bank investments are channelled into co-financing economic infrastructure projects. A similar pattern is also seen from the sectoral breakdown of BADEA financing (see Table 8.4). Co-financing from multilateral, bilateral or private sources help developing countries to supplement limited public sector budgets to improve the financial viability of infrastructure programmes.

Table 8.3   World Bank co-financing by sector, 1991–2000 (percentage of total, by fiscal year)

Sector	1991–2000
Agriculture	7.2
Industrial development	1.1
Economic infrastructure	51.4
Social infrastructure[a]	12.8
Multisector	16.0
Others	11.5

Source: World Bank data.
[a] Includes education, health and nutrition, environment, population and urban projects.

Table 8.4   Sectoral breakdown of BADEA loans from 1975–2001 (US$ million)

Sectoral distribution for loans 1975–2001

Sector	US$ million	%
Infrastructure	875.166	51.52
Agriculture & rural dev't	513.872	30.25
Energy	128.235	7.55
Industry	52.928	3.12
Private	68.190	4.01
Social	47.800	2.81
Special Programme	12.635	0.74
Total	1,698.736	100.00

Source: The Arab Bank for Economic Development in Africa (BADEA).

Co-financing also helps borrowers in developing countries with credit constraints to access international capital markets to increase overall private resources. Although co-financiers may want to reduce their financial risk, the number of co-financiers for each project should be kept to a level commensurate with the complexity and type of the project. Developing a project in partnership with local investors may also help in the reduction of political risks for foreign investors. The joint involvement of local and foreign firms is therefore actively encouraged and targeted by the Asia Infrastructure Fund (IFC, 1996).

Conflicting issues in co-financing

There is now an increasing demand for greater co-operation between development agencies but there are conceptual and operational difficulties associated with the co-financing approach. Conflicts may range from conceptual problems such as definition of sectors and development priorities to operational ones such as an acceptable rate of return and foreign exchange requirement. Commercial sponsors for infrastructure projects rarely consider investments in developing countries attractive unless they have rates of returns above 15%. Sometimes the choice of technology is pre-determined by the investment and appraisal process.

Each funding agency tends to have a set of criteria for funding infrastructure, thereby creating conflicts in the co-financing process with far-reaching implications on preconstruction, construction and operational costs. These difficulties create problems at both ends of the infrastructure delivery process. At the beginning, conflicting interests among financiers during the pre-construction stage add to the complexity of project planning and processing. And at the end, problems of infrastructure sustainability are created as a result of compromises often ignoring the operational implications of infrastructure facilities, domestic needs and development aspirations of developing countries.

Although the Asian Development Bank's consultation with co-financiers starts earlier, BADEA's consultation with other co-financiers, although considered to be one of the most important stages, takes place only after a project is appraised and sent to the Board of Directors for final approval (see Figure 8.10). A more effective co-ordination of funding agencies' programmes will facilitate joint preparation, appraisal and supervision missions, sharing of information, packaging of procurement and harmonising reporting procedures and the timing of disbursements to reduce conflicts and improve co-ordination of infrastructure development.

Lending conditions and sector priorities

Conflicting lending conditions in co-financing has contributed to low and inefficient investment. Lending terms and conditions such as interest rates,

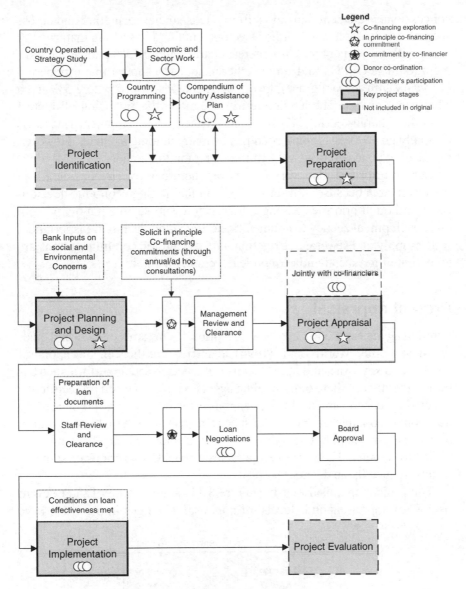

Figure 8.10   Asian Development Bank donor co-ordination and official co-financing process.
Source: Adapted from Asian Development Bank (1997, p. 24).

grace period, maturity, cross default clauses, tied funding and cost-sharing agreements and implementation procedures are strongly influenced by the sources of finance. These can affect the scale and the level of participation in infrastructure development.

Project risk is tied to its core lending conditions and risk exposure declines as the loan is repaid but the rate of decline depends on the repayment terms of the loan agreement. The core lending conditions of a loan agreement are interest rates, grace period and maturity and the degree of

203

concessionary element that varies from one financier to another and across sectors. For example, the Nordic Development Fund has a long maturity (40 years with 10 years grace), no interest charges and 0.75% annual commitment fee. The OPEC Fund, on the other hand, has a shorter maturity period (17 years with 5 years grace), low interest charges from 2% to 2.5% and a 1% service charge. The lending terms also affect the financing of interest and other charges during construction. Priority for development finance is normally given to developing countries by many funding agencies. However, the lending concentrations vary from one region to another.

Sector priorities also vary but infrastructure is often considered an important area because of its critical role in facilitating economic development, although priorities among infrastructure sub-sectors may vary from one development agency to another. Basic education, health, transportation, water supply and sewerage, irrigation, energy and telecommunications are all essential for social and economic development.

Project appraisal

Funding agencies tend to have their own priorities in terms of level and quality of appraisal. World Bank (1996) identified conflicts in the appraisal process as a key issue, and noted that its influence on sector performance has been undermined where other funding agencies have financed projects it had rejected as uneconomical. Project appraisal methodology is dominated by traditional economic techniques (cost benefit analysis, internal rate of return, utility analysis, etc.), but appraisal criteria and priorities vary from one financier to another. There are several relevant fields of appraisal, some of which are new in the debate.

The project appraisal matrix (Figure 8.11) shows some of the relevant 'fields' of appraisals and 'levels' of appraisal. Changes in priorities have

	Levels of Appraisals			
Fields of Appraisal	Target Group	Project Holder or Owner	National Government	Financiers
Economic				
Financial				
Institutional				
Technical				
Environmental				

Figure 8.11   Project appraisal matrix.
Source: Werner (1994).

occurred over the past 10 years. More emphasis is now given to environmental appraisal to reflect the increasing concern of global warming, particularly by the leading industrialised nations. Some funding agencies and countries adopt rigorous and sophisticated environmental assessment techniques whilst others are confined to simple assessments. There are, therefore, likely to be conflicts between donors and host country environmental requirements.

Institutional appraisal has become one of the important dimensions, as the outcome of projects depend on the quality of institutions responsible. The financial aspects of a project have always been a crucial factor but emphasis at the appraisal stage has been too much on initial capital at the expense of recurrent financing and the entire project cycle. Financial appraisal in developing countries is undoubtedly complicated for project holders where there is a mix bag of co-financiers often with conflicting lending terms and conditions.

Cost-sharing and foreign currency constraints

Cost-sharing agreements are widely used in infrastructure development because of limits imposed on financial contributions by development agencies. Cost sharing is the percentage of individual financiers' loans relative to the total cost of the project, excluding taxes and duties. Cost-sharing limits may apply to individual infrastructure projects as well as the overall lending programme of development agencies. The cost sharing limits vary from one financier to another, and across sectors. Some financiers insist on local counterpart funds as a mandatory requirement to raise additional local funding. For example, the World Bank generally expects borrowers to provide a minimum of 10% contribution (local counterpart funds) to project cost while BADEA's contribution does not normally exceed 50% but could reach a maximum of 80% where a project's total cost does not exceed US$ 10 million. IFC funding for private projects is usually limited to 25% of the total project cost in order to attract other private investors.

Development finance also distinguishes between local and foreign exchange components. Most development agencies normally finance foreign exchange costs, but local costs are expected to be met by local counterpart contributions. A key factor taken into account by the World Bank in determining a project's foreign exchange cost is the likely mix of local and foreign awards for a project. Such an approach is problematic, as project holders in countries with foreign exchange constraints may tend to favour a higher proportion of foreign exchange component, which implies a higher participation of foreign firms, at the expense of local participation and resource development.

Foreign exchange scarcity and constraints has contributed to the intermittent supply of vital construction resources in developing countries (Ofori, 1991).

Cross default and other conditional clauses

Co-financing may involve situations where commercial lending is made with cross default clauses relating to loans from official sources. In such circumstances, default on the latter is taken as default on the commercial loans. Some financiers also have conditional clauses in their loan agreements that provide them with the right to cancel the loan agreement if the borrower defaults in paying the interest or other fees within a certain period for other projects they are funding. Under some lending agreements borrowers or project holders may also be required to settle their part of the bill or local cost payments before other financiers can disburse their share of the foreign cost. This requirement is quite common in loan agreements and can be a source of payment delays for consultants and contractors. Bilateral lending agencies sometimes interfere with the procurement process when funding is conditional or tied to the supply of equipment or firms from the donor country, even though the supplier or firms may not always be the best. Two opposing schools of thought exist for bilateral assistance – 'the moralist' and the 'nationalist' (Marsh, 1987). Moralist believes that bilateral finance should be spent without regard to the interests of the industries in the donor country whereas nationalists argue for a maximum return to the donor country's industry. Some development agencies notably, the World Bank, do not administer tied co-financing funds.

Other funding considerations

No discussion on funding infrastructure projects in developing countries will be complete without an understanding of the impact of project identification capacity and risks. Limited project identification and development capacity is a serious factor affecting the development of infrastructure in developing countries (World Bank, 2002). A well-structured project striking an appropriate balance between debt and equity finance, public and private finance is essential in attracting funding agencies and investors.

This process of project identification and structuring is becoming increasingly complex, particularly for infrastructure projects with long payback periods and, high risks. Identifying and understanding risks is crucial in the viability of infrastructure projects as it affects both the costs and revenues associated with any project.

Project identification capacity

There is a lack of project identification and structuring expertise in both the public and private sectors in many developing countries. A dominant public sector that is inefficient in project identification and development reduces the opportunities to attract investment in an ever increasingly competitive financial environment. As a result, many developing countries have had to rely increasingly on the initiatives and project expertise of multilateral and bilateral development agencies during sectoral review missions. Capacity building in project identification and structuring for both the private and public sector institutions is required if the challenges in securing funding infrastructure development are to be met.

Baum and Tolbert (1985) emphasized the need to strengthen the project appraisal capacity of developing countries in order to be able to identify and screen out non-fundable projects with low rates of return, at the earliest stage. Institutional strengthening of projects in the public and private sectors of developing countries, by the development of project capacity is now a key priority of the World Bank and other development agencies. The aim is partly to develop project identification and the preparation capacity of central and local government, as well as other beneficiaries at the community level (Pean and Watson, 1993).

The recent trend towards infrastructure privatization in developing countries is also aimed at improving project identification and development capacity in the private sectors to attract more investment. Capacity for project identification and preparation is crucial as projects to be funded by investors or donors are often scrutinized using many criteria, including economic returns, consistency with sectoral policy priorities and country strategy, and to ensure that the institutional framework is conducive for project sustainability.

Unless project identification and development capacities are sufficiently improved, developing countries will remain uncompetitive, unattractive to bilateral, multilateral and other international investors, particularly private sector investors. Inappropriate design of infrastructure at project conception is often due to the lack of local project development expertise. In many instances, this means that infrastructure projects are not well structured for funding and cases for alternative and more appropriate design and technology to create local employment and to nurture local capacity are not often adequately argued. Technology transfer is often seen as a short cut for enhancing the expertise and capacity of developing countries. However, there are serious problems associated with technology transfer as developed nations tend to see it as a loss of business opportunity whilst developing countries consider it as problamatic due to the risks associated with implementation, usage and long-term sustainability (Carrillo, 1996).

Inappropriate design and the indiscriminate application of equipment-intensive construction technology have reduced the opportunities for local capacity building and poverty reduction in developing countries (Pean and Watson, 1993). Limited project planning capacity means increasing reliance on foreign planning and design solutions. Foreign design solutions tend to favour the use of foreign materials and equipment-intensive technology which affects the sustainability of infrastructure projects. For example, rigid design specifications on compaction standards, thickness of road base and sub-base, and quality of materials has been identified as a typical factor affecting local participation in infrastructure development (Edmonds and de Veen, 1992).

A more flexible approach is required that reflects local resource considerations and procurement systems that would best serve the needs of

207

developing countries in terms of developing and sustaining local capacity (McCutcheon, 1995 and 2001).

It is therefore pointless to talk about appropriate design and technology without addressing issues of developing project capacity to plan well-structured infrastructure projects that can attract funding from development agencies and the private sector.

Risk analysis and management

Analysis and management of risks is central to project finance, particularly for infrastructure projects in developing countries. Risk affects both the outcome of projects in terms of costs and revenues, and therefore project viability. There are different types of risk all affecting to varying degree the different stages of infrastructure development from planning, design, construction to operational stages. There is a need for a comprehensive identification of the different types of risks and the development of a risk matrix and a management plan. The risk profile of a project will affect the funding options as well as the level of funding required. Examples of different types of risks and risk instruments for managing them are shown in Tables 8.5 and 8.6. There are many other risks to be considered such as policy and legislative risks, planning risks, technology risks, inflation risks etc. Some risks are negotiable or transferable while others are not.

Table 8.5   Examples of risks at the design and construction stage

Risks	Risk instruments
Design	Professional indemnity
Cost overruns	Standby credit/contingency funds
Delays	Liquidated damages
Payment defaults	Contractual arrangements
Hidden defects	Performance bonds
Force majeure	Insurance

Table 8.6   Examples of risks at the operational stage

Risks	Risk instruments
Operating efficiency agreement	Service performance
Increase in operations and maintenance	Life cycle/contingency reserve
Market demand	Contractual arrangements
Input availability	Supply agreement
Foreign currency	Central bank assurance
Debt burden	Debt service reserve

A comprehensive risk matrix will enable risks to be allocated to the parties best placed or able to manage them, whether it is a private or public organisation, local or central government, or a local or foreign company. The matrix will also provide an estimate of the probability of the risk event occurring, the cost should the event occur, and the identification of appropriate risk instruments to manage the risks. Adequate costing of risks is important to determine the level of funding required for any infrastructure project. Some of the techniques for assessing and managing project risks are discussed with examples in the previous chapter (Chapter 7).

Case studies

Case study 8.1: Kali Gandaki 'A' hydroelectric project, Nepal

The mountainous topography of Nepal, together with its network of rivers have the potential to exploit hydroelectric power as a major renewable source of energy for domestic, industrial and commercial purposes. The demand for energy in Nepal presents an opportunity for the Nepal Electricity Authority (NEA) to improve its cost recovery by means of enhanced retail tariffs that promote efficient and sustainable power consumption.

The Kail Gandaki River is U-shaped for more than 45 km and is 6 km wide between Mirmi and Beltari in central Nepal. Over this distance, the river drops 108 m and a dam has been constructed at Mirmi to divert some of the water into a tunnel leading to a surface power station at Beltari. There is sufficient water behind the dam to operate the 144 MW turbines at maximum capacity for at least 6 hours per day through the year.

The prime components of the project are:

1. A concrete gravity dam 44 m high complete with a gated spillway and an adjacent intake and de-sanding basin.
2. 5.9 km of concrete lined headrace tunnel leading to the power station, including ancillary equipment.
3. Surge and pressure shaft tunnels leading to the power station, including ancillary equipment.
4. 3No. 48 MW turbines together with controls and ancillary equipment.
5. 2No. 132 KV transmission lines delivering power to Pokhara and Butwal (105.7 km).

The total estimated cost of the project was US$ 452.8 million and the financial plan is outlined below, together with the actual costs determined after the project was completed in December 2003.

From inception the project was completed in 89 months and as shown below the actual project cost of US$ 354.8 million was significantly under the estimate.

Cost	Appraisal estimate	Actual
Project cost (US$ million)		
Foreign exchange	320.0	241.3
Local currency cost	132.8	113.5
Total	452.8	354.8
Financial plan (US$ million)		
Implementation costs		
Borrower financed	83.80	36.30
AsDB financed	156.05	137.90
JBIC	156.05	98.90
Total	395.90	273.10
Interest and service charges (during construction)		
Borrower financed	49.00	77.20
AsDB financed	3.95	2.80
JBIC	3.95	1.70
Total	56.90	81.70
Grand total	452.80	354.80

Legend: AsDB – Asia development Bank, JBIC – Japan Bank for International Co-operation.
Source: AsDB & Nepal Electricity Authority.

The contractor for civil works was Impregilo S.P.A., Italy and the equipment suppliers included companies from the USA, Europe, China and Nepal. Consulting services were provided by Morrison Knudsen, USA and a team of consultants and experts.

Based on average tariffs for residential (US$ 0.090/kWh), industrial (US$ 0.120/kWh) and commercial (US$ 0.153/kWh) and a 50 year project period, sustainability was assessed by comparing the weighted average cost of capital (WACC) to the financial internal rate of return (FIRR). The FIRR for the project was estimated at 12.6% against a WACC of 5.4%, hence, the project is considered to be financially viable.

Author's commentary
The Kali Gandaki hydroelectric power plant demonstrates the effectiveness of co-financing arrangements, where in this case, the main borrowing risk was shared by the AsDB and the JBIC. The project

*was successfully completed but not without difficulty caused by con-
tractual disputes, various strikes and civil unrest in the contractor's
camp. Financial assessment indicates that the project is viable and
sustainable based on the revenue it will generate over a projected
period of 50 years from the completion of the project. The operation
and management of the dam and power station will be the responsibility of the Nepal Electricity Authority.*

Case study 8.2: ROADSIP – Road rehabilitation and maintenance, Zambia

In 2002, Zambia's road network consisted of approximately 67 000 km
of roads, 60% of which was classified. Of the 40 000 km of classified
roads, 7250 km was paved of which only 57% was in good or fair condition. The remaining 32 750 km comprised of unpaved road consisting of an earth and gravel base. Only 7% of unpaved roads were
judged to be in good condition.

The overall poor state of Zambia's roads was and still is a source of
prime concern to the government because of the constraints being
imposed on economic growth and the limitations placed on the mobil-
ity of the working population. The only source of revenue for the Road
Fund intended for repair and maintenance has been wholly dependent
on income from a fixed fuel levy of 15%. In 2002, the levy was equivalent to US$ 0.05 per litre and as a consequence funds generated were
insufficient to cover the cost of repairing and maintaining roads in their
current condition. The key issues concerning road infrastructure in
Zambia were:

- to secure adequate funds for rehabilitation/upgrade, periodic
  and routine maintenance;
- the establishment of agencies to independently manage a
  freestanding system for road fund management;
- the development of resourcing priorities and their resulting allocation;
- improvements in value for money and efficiency; and
- Improvements to road safety.

Under the Road Sector Investment Programme (ROADSIP), targets have been laid down whereby classified paved roads in good
and fair condition will be progressively increased between 2002 and
2006 by 8%. Similarly, the classified unpaved roads in good and fair
condition will be increased by 85% over the same period. The total
length of classified roads will remain unchanged over this period.

The estimates shown below indicate that revenues are totally
inadequate to achieve laid down targets and the only hope is to
attempt to close this gap by donor participation. So far, the IDA has

committed financing alongside the Danish Development Agency, EU, Norway Development Agency, Nordic Development Fund and, hopefully, other donors will also be committed i.e. AfDB, JICA, Kuwait Fund and USAID. It is anticipated that donor financing will amount annually to 46, 65 and 80 US$ millions between 2004 and 2006. This will still leave a shortfall of US$25 million in 2004, US$ 16 million in 2005 and a surplus of US$30 million in 2006, thus producing an overall shortfall of US$ 11 million.

Revenue/expenditure US$ million	2004	2005	2006
Income from domestic funding i.e. fuel levy, user charges and GRZ budget	51	58	70
Estimated fund utilisation			
Civil works, i.e. maintenance, upgrades and repairs, etc.	90	104	90
Engineering and technical services	12	13	11
Capacity building and reforms	16	17	15
Accessibility and mobility improvements	4	5	4
Total	122	139	120
Shortfall	71	81	50

Source: AfDB.

Author's commentary

This case study demonstrates how difficult it is for poorer developing countries to maintain and repair existing infrastructure, given the weak state of their economies. Without intervention and support from donor agencies and partners, sustainable solutions to maintain existing infrastructure are only a remote possibility. Where GDP per capita is very low, i.e. below US$ 850 per annum it is difficult to increase taxation and to make users pay, especially in the case of Zambia where fuel prices are some of the highest in the region. Nevertheless, there is some evidence in this case study that road user taxation is not collected efficiently and accounting systems appear to be less than adequate. There is also a lack of innovation relating to the different ways to raise income by means of realistic user charges. Capacity building is therefore very important to provide the basis for local solutions.

Case study 8.3: Rift Valley water supply and sanitation project, Kenya

Nakuru is Kenya's fourth largest town and it is of economic importance as a major industrial and agricultural centre, as well as having potential in the tourism market. However, the town and its surrounding area suffers from an inadequate and deteriorating water supply. There is also poor institutional capacity to operate, maintain and financially control facilities. In 1993, problems were sufficiently severe to cause an impact on poverty levels and the health of the community began to suffer because of an increase in water-borne diseases. The lack of water was also causing a scaling down of industrial activities with consequential adverse effects on the local economy. In response, the Kenya government (GOK) approached the AfDB to fund a medium term feasibility study into water supply and sanitation, which was completed in 2000.

Following a failed request by GOK in 2001, a revised request for funding was made in 2002 that included other towns and rural areas within the Rift Valley Water Services Board area. Subsequently, this bid was successful in the Spring of 2004 and an AfDB loan was secured amounting to US$ 18.76 million, plus an AfDB grant of US$ 7.22 million was also allocated, which in total represented 85.1% of the total estimated project cost. This was enough to fund all foreign costs and 66% of local costs. GOKs contribution was 14.2% and the beneficiary communities contributed the remaining 0.7% of the total estimated project costs amounting in all to US$ 30.52 million.

The project comprises the following:

- Institute support to the Rift Valley Water Services Board (RVWSB) and the Nakuru Water Supply and Sanitation Services Co. to build capacity in accordance with the Water Act (2002).
- Rehabilitate urban water and sanitation facilities in Nakuru to reduce high levels of waste by leakage and to extend services to the unserved peri-urban areas.
- Address the needs of secondary towns in the RVWSB.
- Determine fulfil the water supply and sanitation needs of the region using a responsive approach.

The outputs are envisaged to be:

1. Effectively functioning water and sewerage service institutions in the project area.
2. Water supply and sanitation systems rehabilitated and augmented.
3. Water and sanitation services extended to low-income peri-urban areas.
4. Better informed community about environmental health and sanitation issues, as well as malaria, HIV/AIDS, gender and other cross cutting issues.
5. Management options and short term investment needs for small Rift Valley town studied and implemented.

213

6. Rural water supply and sanitation programme developed and implemented.

The project's feasibility is judged on the following assumptions:

Financial Internal Rate of Return (FIRR)
A 25 year project life is assumed and the current average tariff of US$ 0.30/m^3 is maintained until the commissioning of the project in 2007. The average tariff will be incrementally increased towards reaching full cost recovery tariffs by 2010 when it will be kept constant at US$ 0.42/m^3.

Operations and maintenance costs are estimated to be 8.5% of capital costs.

The expended losses of water are expected to be reduced from their current level of 70%. In 2005 average losses are expected to be 60% and to stabilise at 25% from 2007 onwards. These figures have been factored into the calculations. The revenue from sewerage is estimated at 14% of water revenues.

The NPV for the project is calculated at US$ 32.32 million against a total project cost of US$ 30.52 million and the FIRR is calculated at 13.86%. Hence, the project is estimated to be financially viable and sustainable based on the above assumptions.

The Economic Rate of Return EIRR additionally takes into account assumed health care benefits and the discount rate has been set at 12%, thus giving an (EIRR) of 24%.

Author's commentary
The Rift Valley Water Supply and Sanitation project has received major funding from the AfDB, without which the whole region's water supply and sanitation would have fallen into terminal decline resulting in serious economic and health problems. This case study underlines the vitally important role fulfilled by the AfDB in providing loan and donor support for infrastructure projects in Africa. Many assumptions have been made regarding the feasibility of completing the project as expressed by FIRR and EIRR calculations. Key to the fulfilment of these expectations is the future economic performance at local and national levels, together with the determination of the indigenous population to drive up living standards and improve the quality of life, given the advantages that this project will bring.

Summary

Events in 2003 have indicated that the wealthiest countries in the world, including the USA are experiencing difficulty in maintaining sufficient investment in infrastructure to guarantee public service expectations. It is therefore not surprising that developing and least well off countries are experiencing difficulties in funding the growing demand for all types of infrastructure. There is a strong relationship between economic growth and the provision of sufficient infrastructure. This is of a cyclical nature because infrastructure development encourages economic growth and a more affluent economy, which in turn makes the provision of additional infrastructure more affordable. Hence, countries able to find a solution to this dilemma take off, while others who are not so lucky flounder in poverty and poor living conditions for inhabitants.

Countries that generate good prospects find it much easier to borrow and they can afford to pay near commercial rates. They also have less difficulty in attracting private investors and foreign inward investment. A nation with a stable economy and strong political leadership with good prospects for growth should have little difficulty in attracting the necessary funds to support policies and goals. The main problem lies with those countries that are genuinely poor and who are unable to draw on natural and human resources to support aspirations for development and growth. Both the multilateral and bilateral agencies have recognised this problem, but there is a danger that some will see the attraction of private investment as a means of abrogating this responsibility. Analysis points to the fact that private investment will not be forthcoming where the conditions are not conducive, as typified by high risk, concessionary interest rates and long maturity periods. World Bank has called for innovative methods to fund such projects, but it is difficult as a private investor to ignore the fundamental economics of investment and the bottom line return. From the statistics presented it is clear that more dependence is being attached to private investment, as in the case of better off developing countries and it is reassuring to observe that most of the multilateral and bilateral agencies are maintaining their funding levels for the least well off nations.

Infrastructure projects invariably require large amounts in investment and it has been shown that the poorest countries can only afford to contribute a small percentage of the equity required. Consequently, advantageous loans are made from multilateral and bilateral agencies, but usually these are heavily conditioned regarding procurement and project monitoring and control. Where expertise does not exist locally to undertake a project, then this will be brought in from outside. In turn this raises issues concerning foreign exchange and sustainability. More attention is being given to technical support and capacity building, such that infrastructure can be properly operated and maintained according to life cycle performance benchmarks.

Due to the size of certain projects, it may be necessary to enter into co-financing arrangements where agencies, private investors and governments jointly contribute to a project. This chapter highlights the conflicts that can occur between co-financing partners due to conflicting interests, it is therefore important that the mix and number of partners is adequately considered before entering into such arrangements.

The importance of appropriateness of design and the innovative use of indigenous resources is a vital consideration in the way that projects are realised for the benefit of those for whom the infrastructure is intended to serve. A major problem is that the gap between the wealthiest and the poorest countries is increasing and there is an inherent responsibility of the advanced industrial nations to assist those countries that have fallen so far behind that they can no longer assist themselves. This is an all embracing problem that requires integration and joint effort from all nations on a global scale. During the early part of this millennium, multilateral and bilateral agencies will face a huge task to provide funding for infrastructure that will truly assist the poorest of the world's nations to drag themselves out of poverty and to begin the long haul to provide the most basic of acceptable living standards for inhabitants.

References

Ahluwalia, M.S. (1997), *'Financing private infrastructure: lessons from India'*, In H. Kohli, A. Mody and M. Walton (eds.), Choices for efficient private provision of infrastructure in East Asia, World Bank, Washington, pp. 85–104.

Baum, W.C. and Tolbert, S.M. (1985), *'Investing in development: Lessons of World Bank experience'*, Finance and Development, December, 26–36.

Carrillo, P. (1996), *'Technology transfer on joint venture projects in developing countries'*, Construction Management and Economics, 14, 45–54.

Daniere, A. (1996), *'Growth, Inequality and Poverty in South-East Asia: The Case of Thailand'*, *Third World Planning Review (TWPR)*, Vol. 18(4), 373–395.

Edmonds, G.A. and de Veen, J.J. (1992), *'A labour-based approach to roads and rural transport in developing countries'*, International Labour Review, 131(1), 95–110.

Gaude, J. and Watzlawick, H. (1992). *'Employment creation and poverty alleviation through labour-intensive public works in least developed countries'*, International Labour Review, 131(1), 3–18.

International Finance Corporation (IFC) (1996), *'Financing Private Infrastructure, Lessons of Experience'*, World Bank, Washington, D.C.

IFC (2003), *'Annual report'*, Available online www.ifc.org/ar2003.

Jinchang, L. (1997), *'Urban employment guidelines: employment-intensive participatory approaches for infrastructure investment'*, [http://www.ilo.org/public/english/125polde/papers/1998/guidelin.htm], October.

Kisanga, A.U. (1996), *'Towards sustainable private sector investment in infrastructure projects in Sub-Saharan Africa'*, African Construction Bulletin, Issue No. 4, December.

Mabogunje, A. (1993), *'Infrastructure: the Crux of Modern Development'*, The Urban Age, (Washington), 1(3), 3.

Marsh, P.D.V. (1987), '*The art of tendering*', Gower Technical Press, Aldershot, England.

McCutcheon, R.T. (1995), '*Labour-intensive construction in Sub-Saharan Africa: the implications for South Africa,*' Habitat International, 19(3), 331–355.

McCutcheon, R.T. (2001), '*Employment generation in public works: recent South African experince*', Construction Management and Economics, 19, 275–284.

Ofori, G. (1991), '*Programmes for improving the performance of contracting firms in developing countries: a review of approaches and appropriate options*', Construction Management and Economics, 9, 19–38.

Pean, L. and Watson, P. (1993), '*Promotion of small-scale enterprises in senegal's building and construction sector: the 'AGETIP' experience,*' In New Directions in Donor Assistance to Micro-enterprises, Organisation for Economic Co-operation and Development, Paris.

Sethuraman, S.V. (1997), '*Urban Poverty and the Informal Sector: A critical Assessment of Current Strategies*', International Labour Organisation, Geneva.

Wells, J. (1993), '*Appropriate building technologies: an appraisal based on case studies of building projects in Senegal and Kenya*', Construction Management and Economics, 11, 203–217.

World Bank (1996), '*Guidelines: procurement under IBRD loans and IDA credits*', World Bank, Washington.

World Bank (1988), '*Road Deterioration in Developing Countries: Causes and Remedies*', A World Bank Policy Document, World Bank, Washington, 1–12 World Bank, Washington.

World Bank (1998), '*$309 Million for Ethiopia's Roads: World Bank's Largest-Ever IDA Credit to Africa,*' [http://www.worldbank.org/html/extdr/extme/1613.htm], January.

World Bank (2002), '*Private infrastructure: a review of projects with private participation, 1999–2000, Public policy for the private sector*', June.

Part 3

Operational considerations

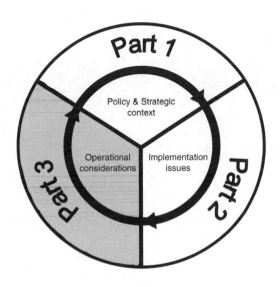

Key themes

sustainable design
conservation
facilities management

Chapter 9

Towards sustainable infrastructure and conservation

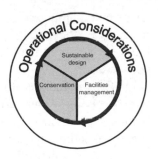

Introduction

It is generally accepted that the majority of existing infrastructure throughout the world is not sustainable in the long term and there is an increasingly urgent need to implement changes in strategic provision. Design choices, construction methods and operations are central to providing sustainable infrastructure for the built environment to satisfy the need of governments for human development and economic growth.

Infrastructure plays a central and vital role in the well-being and prosperity of society and it is therefore important to understand all aspects of its relationship to sustainability and conservation. Infrastructure and economic growth are intrinsically linked. Sustainable development is also dependent on economic growth but this will only be achieved by improvement in the design, construction and operation of infrastructure facilities. Greater efficiencies created by sustainable infrastructure will lead to reductions in waste, energy consumption and pollution together with improved utilisation of space and minimisation in the consumption of non-replaceable and replaceable resources. These goals have far-reaching implications that affect every aspect of society and they have a fundamental influence on strategies relating to social planning and urban and rural development.

Guiding principles and approaches to sustainability and climate change have been agreed by nations at the Rio Earth Summit in 1992. These principles have subsequently been confirmed and enhanced by further summits held in Japan and South Africa. To achieve targets relating to reductions in environmental pollution and the arrest of resource depletion, a concerted international effort is required by all nations to avoid the consequences of global pollution, shortages in non-replaceable resources and the potential havoc that might occur because of climate change. The role of infrastructure in terms of the way they are designed and constructed to operate in the built environment is central to achieving the goals of sustainability and conservation.

Nations committed to implementing sustainable infrastructure policies will be taking a long-term view to achieve higher economic growth and human development through greater efficiency in energy consumption, space utilisation and conservation, and recycling of replaceable and non-replaceable resources. This also implies research and development in the use of alternative fuels and energy sources for the operation of infrastructure, e.g. hydrogen and solar energy. By definition, if this initiative, along with the development of new design and technologies are successful, then those countries that have made this investment will be far better positioned than those nations that have chosen to ignore the call for sustainable and conservation policies in the operation of infrastructure.

There is an increasing awareness that the capital cost of providing infrastructure represents only about 10% of the full life cycle cost and therefore the potential to gain most efficiency saving derives from a long-term operational policy framework, underpinned by efficient design and construction, together with well-managed infrastructure facilities. There will also be the benefits to the users who will be able to take advantage of more efficient services used in the commercial and domestic environments. In turn, this will promote stronger economic growth, better living conditions and a robust built environment as a result of more sustainable infrastructure, better conservation and efficient space utilisation. Further, benefits will be gained by means of less pollution and more affordable public services.

Conservation of urban and rural environments by means of sustainable infrastructure development is crucial to a well-balanced and structured society that can enjoy high-quality townscape and take advantage of leisure provided by open space and natural countryside.

This chapter begins by explaining the concept of sustainable development, and the guiding principles derived from recent earth summits. Approaches are explored as to how these concepts and principles can be operationalised with regard to physical infrastructure. Consideration is given to the main drivers that influence the development of an operational plan for delivering infrastructure objectives in support of sustainable development and conservation, and the necessity to improve the human living and built environment. The impact of selected infrastructure systems on sustainability and conservation is discussed. Recommendations are made concerning the use of various assessment methodologies for monitoring the environmental performance of infrastructure with a view to recognising the need for further change and timely development, thus closing the regeneration loop.

Definition of sustainable development

At the World Summit on Sustainable Development held in Johannesburg in 2002 world leaders agreed to commit themselves according to the following:

> *'Encourage and promote the development of a 10 year framework of pro-grammes in support of regional and national initiatives to accelerate the shift towards sustainable consumption and production to promote social and economic development within the capacity of ecosystems by addressing and where appropriate de-linking economic growth and environmental degradation through improving efficiency and sustainability in the use of resources and production processes, and reduce resource degradation, pollution and waste'*

Sustainability may be defined as the 'capacity for continuance into the long term future' (Forum for the Future, 1999). Anything that can continue indefinitely is sustainable. Sustainable development has been defined by Brundtland (1987) as 'Development that meets the needs of the present without compromising the ability of future generations to meet their own needs'.

Models of sustainable development

Sustainable development comprises of a complex interaction of three dimensions, namely environmental, economic and social. Most people have little difficulty in understanding the environmental dimension that concerns the efficient use of resources, recycling, conservation, energy consumption and climate change. However, the economic and social dimensions are less tangible and more difficult to understand and quantify. The complex linkages that play a vital role in the holistic nature of sustainable development compound this lack understanding. Figure 9.1 illustrates this interaction.

A major problem has been the perceived need to generate economic growth through increased infrastructure investment to meet demands from business and society. This in turn has impacted on the environment by creating unsustainable use of non-replaceable resources, excessive waste

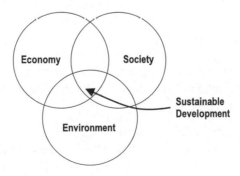

Figure 9.1   Triple link sustainability model.

and pollution as a result of inappropriate infrastructure design and construction processes. More significantly, the impact is greatest at the operational stages due to the long-term nature of their use. It is now widely recognised that the release of greenhouse gases (GHGs) into the atmosphere is helping to cause global warming and climate change. Furthermore, natural material and energy sources will at some time in the future become fully depleted at the current rate of consumption. Sustainable infrastructure design and conservation will therefore help to defer the time of depletion and more importantly will give additional time to find and develop alternative sources that are either abundant or replaceable.

Previous studies (Ekins et al., 1992; Serageldin and Stee, 1994) have further conceptualised the Triple Link Sustainable model by the introduction of a 'Five Capital Model' that represents all the resources available to a society for the achievement of sustainable development. Each form of capital is developed as follows:

1. *Natural Capital* – represents the stock of assets presented by the environment that can be categorised as renewable, e.g. trees, vegetation, water and non-replaceable, e.g. fossil fuels and minerals. There is also the possibility that renewable resources may become non-replaceable due to climate change.
2. *Human Capital* – represents all those things that contribute to the quality of life and well-being, e.g. knowledge, skill, health, morale, motivation and spiritual comfort. It is considered that investment in the creation of human capital can provide high returns, especially in developing societies.
3. *Social Capital* – represents the fabric of society that can be defined as the bodies and institutions that form a framework to facilitate the way people work and live together. These, for example, may be governments, businesses, universities, charities or families and they all interact in a vast variety of ways in contributing to social capital.
4. *Manufactured Capital* – represents the entire infrastructure that has been created to support the quality of life and well-being, e.g. roads, bridges, power stations, buildings, equipment and machinery. It should be noted that manufactured capital does not include goods and services.
5. *Financial Capital* – represents the productive strength of all other types of capital and thus enables them to be owned or traded. Therefore, financial capital acts as an intermediary for the purpose of trade and commerce.

The balance of investment in the various types of capital will determine the nature and state of individual societies. This will be determined by values, relativities and priorities, which to some extent helps to explain the differences between societies. However, the two primary forms of capital that determine wealth are Natural Capital and Human Capital. It can therefore

224

be stated that the creation of wealth may not always accord with the development of sustainability, especially where non-replaceable resources are used irresponsibly. This issue resides at the core of the sustainability argument and there is ample evidence in the past and present of wealth creation in the form of economic growth overriding all else.

Infrastructure is needed to facilitate economic growth but it is now recognised that a more responsible approach should be taken to decouple economic growth from wealth creation and well-being by the introduction of emphasis on social and environmental integrity. Using the InfORMED approach developed in chapter 5, infrastructure projects can be evaluated and implemented to enhance the development of economic, social, environmental and institutional performance and the various forms of capital. The use of the InfORMED model is therefore consistent with achieving the goals of sustainable development underpinned by the Triple Link Sustainability and Five Capital Models. This means a greater need for understanding the impact of design decisions on the construction and operation of infrastructure. Within this context sustainable construction has been defined (Kibert, 1994) as the creation and responsible management of a healthy built environment based on the prudent use of resources and ecological principles.

Why sustainable design and construction?

Since the summit on Environment and Development in 1972 the weight of environmental evidence has steadily grown to support the case that unprecedented levels of economic growth and affluence are creating climate change, inequality and human injustices. These were comprehensively documented at the Johannesburg Summit in 2002.

Advances in technology have enabled assessments to be made regarding deforestation, coastal erosion, pollution and the presence of GHGs. Global warming and climatic disturbances are cited as being caused by pollution created by GHGs.

The impact on the wholesale use of non-replaceable natural resources and the consequences of not managing the use of replaceable resources, e.g. softwood timber are being more accurately assessed (Parkin, 2000). For example, according to Parkin et al. (2003), the UK construction industry

- consumes 6 tonnes of material per person;
- freshly quarries 80% of its aggregates (only 20% being recycled);
- generates 70 million tonnes of waste per annum;
- throws away 13 million tonnes of unused material delivered to sites per annum; and
- is the most frequent industrial polluter, with increasing accidents against a falling industrial trend.

The design of sustainable infrastructure and conservation is therefore central to reversing this trend. In recent years the environmental, economic

225

and social impact of large infrastructure projects has been subjected to increasing critical scrutiny. The effect of large dams, expressways, airports, power stations and waste disposal plants have been singled out for special treatment that has highlighted the conflict between economic growth and the consequential detrimental effects on society (i.e. the quality of life) and the environment. Decisions concerning major infrastructure projects should as a matter of course be subjected to ethical and value-driven processes from the outset to enable conflicts to be properly surfaced and evaluated. Where needs are justified then ways should be found that mitigate adverse environmental and social effects (ICE, 2004).

The imperatives to be addressed within the context of sustainable construction include:

- Design concepts, including materials/technology selection
- Construction and demolition techniques
- Whole life cycle performance and costing
- Procurement
- Site planning and management
- Recycling
- Waste and energy minimisation.

The implication is that sustainability should be embodied as a primary element of the design process and should be reflected throughout the whole life cycle. Outcomes should be monitored using 'Key Performance Indicators' (KPIs) for both design and construction activities and feedback systems should be established to institute cycles of continuous improvement.

Strategy for sustainability and conservation

Infrastructure is at the heart of sustainable development and conservation as it affects urban, semi-urban and rural environments. There is a need for the establishment of a robust and realistic strategy, linked to an operational framework. As such this should be based on existing and future needs and should holistically address all relevant social, economic and environmental imperatives with a view to encouraging sustainable solutions while conserving resources and heritage. The specific strategy for sustainability and conservation should evolve within the overall policy and strategic frameworks discussed in Chapters 3 and 4 and will cover short, medium and long-term timescales up to 50 years.

The strategy at the highest level will represent a national plan within which regional and country plans will be incorporated. These will reach down to local communities to create an integrated and coherent strategy at all levels as shown in Figure 9.2.

The strategic plan will be linked to an operational plan as the means for implementing and controlling development at all levels (Pitts, 2004).

Figure 9.2   Strategic planning framework.

Moreover, it is essential that sufficient analysis is undertaken to predict economic and technological trends so that realistic assessments can be made to determine what needs to be done to ensure sustainable development. This implies that a thorough understanding exists of how the current position has been achieved. Using this understanding, goals and targets for all aspects of infrastructure development, particularly the operational stages, can be determined to achieve solutions that are sustainable.

By definition, the built environment incorporates infrastructure – civil engineering structures and buildings that combine to form the basis for the existence of urban, semi-urban and rural societies. The current position determines the nature, size and disposition of cities, towns and villages whose existence has been brought about, in some cases, by centuries of development and redevelopment. Many cities and towns owe their development to past times when imperatives such as location, demography and proximity to natural resources determined the form and scale of development. Compared with today, past economic, social and environmental solutions may have been based on comparatively low levels of growth, low technology and different social attitudes and values. As time progressed, there have been development and change resulting in areas of decay and growth, thus triggering new development, redevelopment and regeneration.

The continuing growth of commerce and trade traditionally casts a complex web of conflicting demands on resources at all levels of society. This has manifested itself in transport networks, power generation and distribution networks, water treatment and distribution systems, sewage treatment, waste disposal and communication systems designed to sustain cities and towns. The same but more limited services have been provided to support rural communities. The addition of hospitals, educational establishments, public buildings, housing and special purpose buildings creates a complex built environment that is difficult to effectively control because of conflicts brought about by what already exists and the need for continuing change. The implication is that the steady-state perfect built environment does not exist; nevertheless, it is vital that there should be constant effort to

227

bring about improvement using a holistic and integrated approach. It is now widely recognised (CIC, 2004) that cities and towns of the early 21st century are not sustainable because of their overdependence on the consumption of non-replaceable material resources together with insufficient recycling and waste prevention. Hence, it is necessary for strategic and operational plans at all levels to incorporate sustainable solutions to conserve non-replaceable resources and to divert usage to replaceable resources where possible (Griffiths et al., 2003). This requires a serious approach to regeneration by upgrading and refurbishing suitable existing infrastructure, rather than demolishing and redeveloping.

In future new development should be more rigorously justified, especially with regard to size, diversity and replacement. To realise sustainable long-term development existing planning processes will need to become more sophisticated, where national, regional and local plans are incorporated within a single strategic framework, as shown in Figure 9.3, linked to an operational plan aimed at monitoring and improving the environmental performance of infrastructure. This implies a detailed and knowledgeable understanding of how the economy works at all levels taking into account resource dispersion, ecological, topological and natural environmental issues. The aim should be to develop communities of optimal size that reduce unnecessary journeys from home to the work place and are able to sustain themselves, as far as possible, by the use of local resources, especially those that are from replaceable sources.

Figure 9.3 Integrating urban and rural planning systems within the strategic plan.

228

Operational considerations for sustainability and conservation

There are a number of factors that can be considered in the context of infrastructure development to achieve the objectives of sustainability and conservation. This will enable effective monitoring of the operational framework linked to the strategy for continuous improvement as outlined in the previous section. This includes monitoring climate change and global warming, encouraging a low-carbon economy, zoning of land to improve the built environment, examining the impact of key and large physical infrastructure networks such as transport, energy and water systems, reviewing waste management and disposal systems, and using assessment methodologies to determine or predict the environmental performance by the use of key indicators.

Climate change and global warming

Fundamental to sustainability policies is the need to address issues that directly impact on climate change. This has been recognised since the 1980s when scientific consensus established that climate change was a major global issue. An International Panel on Climate Change (IPCC) was established in 1988 and from this emerged the UN framework Convention on Climate Change, which was signed at the Rio Earth Summit in 1992. From this development the following basic principles emerged:

- Outstanding scientific uncertainties should not be used as an excuse for inaction.
- Action should aim to stabilise greenhouse gas concentrations at safe levels.
- Such actions should be based on common but differentiated responsibilities between countries and that industrialised countries should take the lead in resolving the problem.

The initial objective was for industrialised countries to reduce their emissions to 1990 levels. To help oversee this the Conference of Parties was established.

The Kyoto Protocol adopted in 1997 laid down legally binding targets for each developed country to reduce GHG Emissions by 2008–2012 using 1990 as the base year. However, Australia and the USA refused to ratify the Protocol and only recently it has been ratified by Russia after years of procrastination. Table 9.1 shows the targets for GHG Emissions.

Under the EU Emission Trading Scheme, the first phase 2005–2007 is for carbon dioxide (CO_2) with an expansion to cover all GHGs by 2008–2012. The first phase covers approximately 45% of EU CO_2 emissions from combustion installations with a rated thermal input exceeding 20 MW involving mineral oil refineries, coke ovens, production and processing of

Table 9.1   Targets (2008/2012) Kyoto Protocol

Country group	Target (%) (base 1990)
Switzerland, EU, Central and East European States	−8
Japan, Canada, Hungary	−6
Russian Federation and Ukraine	+0
Norway	+1
Australia	+8
Iceland	+10

ferrous metals, mineral industry including cement and clinker, glass and ceramic bricks, pulp paper and board activities.

Organisations required to meet targets will be allowed to do so by

1. reducing own emissions;
2. reducing own emissions below target and sell or bank the excess emission allowances; or
3. let emissions remain above target and buy emission allowances from other participants.

There are six GHGs, the most prominent being CO_2. The other gases are methane, nitrous oxide, hydro fluorocarbons and sulphur hexafluoride. The following measures are proposed to reduce the cost of compliance:

- Carbon emissions can be offset by 'carbon sinks' such as forests that absorb carbon back from the atmosphere.
- Permission to take credits from emission reducing projects in other countries by means of the Clean Development Mechanism (CDM) or Joint Implementation in developing countries.
- Facilitation of an international trade in credits.

In order for the Kyoto Protocol to come into force, ratification is required from parties to the Convention, who are responsible for 55% of industrialised countries CO_2 emissions in 1990. As at the end of November 2003, 120 countries had ratified the Protocol, but these only represented 44.2% of CO_2 emissions. This has been boosted by Russia's ratification of the Protocol, which can now be enforced.

The EU has taken a positive stance on reducing GHG emissions, which is in contrast to the USA who has refused to date to ratify the Protocol. The EU signed the Protocol in 1998 under which 15 Member States were assigned single emission targets representing an average reduction of 8% over 2008–2012 from the 1990 baseline as shown in Figure 9.4.

For these proposals to work effectively a scheme will need to be introduced that will impose a climate change levy to encourage improved

EU Member State	%
Austria	−13.0
Belgium	−7.5
Denmark	−21.0
Finland	0.0
France	0.0
Germany	−21.0
Greece	+25.0
Ireland	+13.0
Italy	−6.5
Luxembourg	−28.0
Netherlands	−6.0
Portugal	+27.0
Spain	+15.0
Sweden	+4.0
United Kingdom	−12.5
EU 15	−8.0

Figure 9.4   GHG reduction target EU burden sharing agreement (1990–2008/2012).

energy efficiency and reduce GHG emissions. The rates applied in the UK for example are:

- £0.0015/kWh for gas
- £0.017/kg for coal
- £0.0096/kg for LPG
- £0.0043/kWh for electricity.

Note: The levy does not apply to domestic or transport sector fuels or fuels used for the production of other forms of energy (e.g. electricity generation).

There should also be in place a system of exemptions that will take into account where it is sensible, or in the national interest to grant temporary or more permanent exemptions.

Towards a low-carbon economy

At the heart of the barrier to achieving a low-carbon economy is the linkage between growth and increasing energy use by business and the broader economy. To break this connection new approaches must be introduced based on new technologies that reduce energy consumption and place emphasis on establishing energy generation using renewable resources such as wind power, solar energy and hydroelectric power, including wave and tidal flow generators. There is also the possibility of making use of energy generation from hydrogen as a fuel.

Alongside renewable energy, technological development is aimed at the use of greater energy efficiency generated from existing replaceable sources by the judicious application of existing technology and conservation methods. It is argued that such reductions will cut emissions, thus mitigating the adverse effects of climate change and risk associated with climatic disturbance, global warming and flooding.

An environmental argument put forward by conservationists is aimed at greater efficiency in energy use and the use of new technologies to provide new opportunities for businesses and national economies. Hence, conservation measures will not necessarily have an adverse affect on national GDPs. Instead economies will be less dependent on depleting resources and greater confidence can be expressed in the use and availability of sustainable sources of energy that will impact on future waste reduction.

Zoning of land

The zoning of land should be dynamically and sympathetically considered, where 'Brown field' sites can be harmoniously redeveloped to improve urban regeneration and the quality of the environment (CIRIA/CIC, 2003a,b). This implies more consideration to good neighbourhood doctrines to prevent nuisance and aggravation caused by noise and environmental pollution.

Key actions to improve the built environment include:

- Proper social, economic and environmental appraisal.
- Involvement of stakeholders.
- Involvement of the local community as proposed by Agenda 21 of the Rio Earth Summit.
- Communities should be planned to anticipate and incorporate future need for expansion and change.
- 'Brown field' sites require proper appraisal, including the need for decontamination, stabilisation and the clearance of structures from previous use.
- The location of residential communities should be made for easy travelling to and from work, together with good integrated public transport.
- Attention should be given to creating green areas and adequate recreational space.
- Strategic plans should ensure the right balance buildings and the necessary infrastructure to support them, including transport and utility services.
- The design and construction of buildings and infrastructure should provide for efficient use of material resources, energy efficiency and recycling.

Impact of physical infrastructure

Transport systems

Advanced industrialised countries are facing the increasing problem of old and deteriorating roads, railways and waterways that are failing to meet the ever increasing demands being made upon them. This is causing congestion on the busiest roads and poor levels of service on the railways, which in turn creates wasted time and delays. Pollution caused by the internal combustion engine and dependence on oil is a major source of concern.

The aim should be to reduce the number of journeys by efficient and effective land use and the use of developments in communication and information technology (CIC, 2003a). There is also a need for the implementation of holistic strategies that integrate different transport systems to provide ease of access, together with regular and reliable services using public transport. In this manner vehicle use for personal needs, supply and distribution of goods and business requirements can be moved away from the roads and onto public transport, e.g. trains, trams and buses.

A major problem for advanced industrialised countries is the provision of new roads and rail tracks to meet increasing levels of demand. There is also the argument that more roads simply encourage greater road use resulting in further increases in congestion and pollution. The upgrading and renovation of existing roads usually means that work must be phased to accommodate continued, albeit depleted use and this in turn causes increased costs, extended construction periods, congestion and delays.

In developing industrialised countries there is a need to build infrastructure to support economic growth and to satisfy demand for greater freedom of movement created by increased prosperity. Therefore new roads, airports, waterways and port facilities increase the potential to contribute to pollution as well as increasing the use of the World's non-replaceable resources. However, there is the opportunity to take a holistic view of sustainable development, whereby strategies can be developed and implemented to maximise freedom movement while reducing unnecessary journeys to avoid wasteful congestion and atmospheric pollution.

In developing countries the main issue is a lack of transport infrastructure that restricts economic growth and prevents easy access to markets for trading goods and services. Weak economies find it difficult to afford adequate transport systems and assistance is sought by means of aid and cheap loans provided by the World Bank and regional development banks. The main problem for poorer developing countries is not pollution, or sustainability, rather it is about the growing gap in prosperity with industrialised countries. The system of carbon credits proposed by the Kyoto Protocol is intended to benefit developing countries in that they can trade these credits with industrial producers who cannot meet their GHG targets.

Despite the need to address major and difficult transport issues there have been successes which have the potential to be further developed. Attempts have been made to integrate public services, whereby rail and bus stations are combined and in theory scheduled timetables are intended to coincide. The introduction of GPS systems to inform passengers regarding the arrival time of buses at bus stops has help to restore passenger confidence in public services by keeping them better informed. Some cities have introduced or extended electric tram and light rail systems that provide clean and efficient transport in central and surround areas. More flexible ticketing arrangements have been introduced to allow travel on all modes of transport, thus providing a level of service that is more likely to persuade motorists to leave their cars at home. To further increase the use of public transport some cities have introduced congestion charging, together with increasing the cost of car parking. On a more general level, some governments have increased fuel tax to deter road use and, hence, reduce the consumption of petrol and diesel fuels.

Transport links between cities have been addressed in many countries by upgrading rail tracks to run high-speed trains. Suburban and commuter services have also been improved. Advanced industrialised countries usually have well-developed arterial expressways that link regions, major cities, towns and the countryside. Some countries have adopted toll systems that generate revenue from expressway use, while others have allowed free use of the expressway network. Experience in France has served to drive traffic onto national routes that are toll free and there is some evidence that regions have become more isolated because of the cost of making long journeys (Farrall, 1999).

Most countries have international and domestic air services supported by major airport hubs, regional and local airports. As a rule airports are well served by transport services to the nearest cities or towns.

Waterways and canals have fallen out of use, with the exception of major rivers where goods are imported or exported through coastal ports and water-borne transport is provided inland by barges. There is a growing trend to use inland waterways and canals for recreational and leisure use. This has led to the regeneration of inland locks, tunnels and aquaducts.

Despite progress to date, much still needs to be achieved. Fossil fuels on which most transport relies will eventually be exhausted and it is imperative that alternative fuels are developed, which use abundant or replaceable resources. Research has resulted in the production of electrically powered vehicles using rechargeable batteries and photovoltaic cells; however, an affordable and practical mass transport fuel solution is still to be found. Research is ongoing into the generation of motive power from the use of hydrogen-based fuels. Investment is required to incrementally improve the performance of urban transport networks and acceptable solutions also need to be found for rural areas with low population densities.

Listed below are notional key actions for the development and improvement of transport provision:

- Consideration of national, regional and local transport policies within an overall transport strategy that fully accounts for demographic, business and technological trends and is geared to reducing unnecessary travelling and haulage.
- Full integration of all transport systems.
- City and townscapes that encourage people to ride bicycles or walk, rather than using private or public transport.
- Investment to improve amount and quality of public transport.

- Means should be explored and implemented to encourage public transport use by means of congestion charging, tolls, fuel taxation and parking charges.
- Flexible and efficient haulage systems should be developed using combinations of road and rail transport.

Finally, attention should be given to the technological design and development of all forms of transport that are more reliable, fuel-efficient and provide increased levels of passenger comfort and safety.

Energy systems

Energy generation is traditionally centred on power stations burning oil, coal, natural gas and nuclear power. Normally, these are located away from urban centres and distribution grids comprising over and under ground power lines and substations ensure that adequate supplies of electricity to meet demand. Considerable variation normally exists between low and peak demand periods and in the event that demand exceeds supply, then as a final resort, parts of the system will be shut down. Careful attention is required regarding the timing of maintenance and repair to minimise the need for load shedding and power cuts.

Nuclear power is relatively clean in terms of atmospheric pollution; however, nuclear reactors generate radioactive waste, which must be contained within the reactor until decommissioning when it is reprocessed. There is always the danger of leaking radio active deposits into the atmosphere by accident and expensive safeguards must be in place to reduce the risk of such events occurring. The most potent waste from reactor cores is sealed in strong lead lined containers and then buried deep underground where the possibility of leakage is extremely unlikely. Where nuclear waste has to be transported for reprocessing then this presents significant risk of contamination and there are moves to scale down and decommission nuclear power plants (Ofgem, 2003).

To mitigate the amount of GHG emissions into the atmosphere there are moves to reduce energy produced by fossil fuel power stations that still produce the majority of energy to satisfy demand. This presents a huge task that requires considerable investment in power generation from alternative sources, e.g. wind, flowing water, solar energy and the incineration of rubbish. Energy generated by wind turbines located on rural sites in remote areas, or off shore, have the prime requirement of exposure to uninterrupted wind force. Rural wind farms typically produce 500 MW of electricity for the national distribution grid from infinite supplies of wind energy. Similarly, hydroelectric power produced by dams across fast flowing rivers are another example of harnessing natural resources that are replaceable. The greatest example of hydroelectric power generation is the Three Gorges Dam across the Yangtze River in central China. In countries with suitable terrain and sufficient rainfall old watermills are being refurbished and brought back into commission to generate energy for local communities and individual homes, primarily in rural areas. Solar energy and heat generation using heat pumps and photovoltaic cells currently produces power, only small-scale power. However, it would appear that potential may exist to produce substantially more energy from this source as the result of future research. The demand for energy consumption can be reduced by the appropriate design and location of buildings to take maximum advantage of solar gain retention of heat, thus reducing energy requirements for heating and hot water (Sustainable Task Group, 2004). This is particularly relevant in climates that receive large quantities of sunshine through the year, since they are less likely to have the advantage of hydroelectric power and the potential of generating wind energy may not be as great. In hot climates, air-conditioning systems consume considerable amounts of power, and the introduction of passive air-conditioning systems that take advantage of natural ventilation assisted by induced temperature variation can significantly reduce the need for the consumption of electricity.

Technology has a key role to play in facilitating more efficient power generation by means of electrical goods and vehicles that consume less energy per unit of output. Smart monitoring systems will also help to reduce waste by automatically scaling down output from equipment and plant that is not being fully utilised. In this area people have a distinct role to play by being more aware of waste generation and applying a thrifty culture of turning off equipment when it is not required.

The increasing use of replaceable energy sources linked to a culture that appreciates the importance of conservation will afford more time to discover and develop alternative abundant clean energy sources. In the meantime the reduction in burning fossil fuels will mitigate the emission of GHGs with consequential effects on global warming. Key actions suggested are as follows:

- Strategies should be in place to balance supply and demand taking into account targets to reduce dependence on fossil fuels and nuclear power by the introduction of greater capacity to produce energy from new energy sources that are renewable and abundant.
- The built environment should be developed to reduce demand for energy by means of improved building designs that make better use of solar gain, passive ventilation; and the use of materials and components that use less embodied energy during manufacture and have better life cycle performance.
- Regional and local energy strategies should be closely geared to national strategies and a holistic approach is required that ensures harmonious effective and efficient land use to assist planned reductions in energy consumption.
- The general public require to be educated about the most effective methods to conserve energy and to reduce demand by changes in attitude and life style.

236

- Innovation and developments in technology aimed at both supply and demand sides requires encouragement and funding to bring forward new ideas and concepts that have the ability to develop new energy sources and to conserve those that already exist.

Water systems

The availability of water primarily depends on the amount and dispersion of annual precipitation, together with topography, geology and degree of vegetation. Growing urban conurbations have placed increased demands on the supply of water, which in arid climates can cause severe problems and shortages. In contrast, parts of the World that receive plentiful well-distributed annual rainfall only experience restrictions in unusual periods of severe drought. More recently, there is evidence that climate change is creating unusual weather patterns that cause storms and heavy rainfall resulting in flooding followed by long dry periods. Under such conditions it is essential to capture rainwater and retain it in reservoirs and catchments in order to create a better balance between supply and demand and to mitigate the dangers and adverse effects of flooding.

Once water is retained it requires treatment to make it potable before distribution by means of a mains network to business, agriculture and domestic users. In dry regions potable water is a relatively scarce resource and water for cleaning and flushing toilets could use recycled grey water of lesser quality. Grey water may also be suitable for irrigating crops and providing water for domestic gardens.

In developed countries urban potable mains water is normally recycled several times by means of sophisticated water treatment processes, but in least developed countries such processing plants are unlikely to be available in many areas due to the amount of investment required. In some urban areas and a large proportion of rural areas it is still the case that water is supplied through communal standpipes or natural wells. Much can be done under these circumstances to persuade people to store roof rainwater and to use recycled grey water.

In extremely dry countries that have sufficient coastline desalination may be a necessity, but the cost of energy required to extract salt from seawater is high and the taste is not so good either. Nevertheless, desalinated water can be used for all purposes by industry, commerce and the domestic population, the exception being use for drinking.

Some industries require large volumes of water and it is usual to site factories and processing plants near to rivers and waterways. Careful control is required over the amount of water that is extracted and post process water may require treatment before it is reintroduced back into watercourses.

The maintenance of water distribution systems is a major issue, since water leakage caused by burst pipes can represent a significant proportion of the total amount of water used. Water loss can be attributed to ageing and corroded iron and steel pipes that eventually fail. More lately, distribution mains have been run in plastic and these are proving to be very durable and flexible.

Water plays a key role in the treatment and disposal of sewage by means of drainage networks that flow into treatment works comprising of anaerobic tanks that separate the sludge and aerobic filters that treat the liquid before it is reintroduced into rivers or the sea. In rural areas it is usual to use cesspits and sceptic tanks that require periodic emptying and disposal to designated sites. The alternative is to use small treatment plants for isolated residential communities.

The key issues are:

- An integrated approach to balancing supply and demand, including the means to retain water for use in periods of low rainfall.
- Policies to conserve water and eliminate unnecessary usage through planned system management, prompt emergency repairs and metering business, agricultural and domestic users.
- Greater use of grey water for non-potable needs.
- Development of rainwater collection systems for urban, semi-urban and rural areas.
- Greater awareness of the effect of urban non-permeable surfaces and their contribution to patterns of surface run off leading to flooding.
- Affordable and efficient sewage treatment systems.

Waste management and disposal systems

Typically, the construction industry accounts for 15% of a developed country's total waste. Most of this occurs through demolition and refurbishment or from new build site waste. The conventional way of disposing of waste is to bury it in designated landfill sites, but there is a growing realisation that waste materials, if properly managed can be recycled to produce new or reprocessed materials, e.g. steel, glass, insulation, etc. Concrete, tarmacadum, bricks, aggregate and other solid inert materials have the potential to be crushed and graded for use as fill to form sub-bases and foundations for built facilities.

When waste is deemed not recyclable then it should be classified to determine the most appropriate means of disposal. Suitable combustible materials may be incinerated in power stations, but the benefits should be balanced against atmospheric pollution and potentially dangerous gases that may escape despite filtration.

The key issues are:

- to design out waste during the construction process, whether it be new build or refurbishment;
- to make provision for economic recycling of materials, components and structural elements;
- to reduce consignment to landfill; and
- to generate energy from the combustion of rubbish.

Assessment methodologies

In order to demonstrate the effectiveness of sustainability and conservation schemes and measures it is necessary to use assessment techniques to accurately gauge performance and thus the benefits to be derived. There is no single assessment system that will suit all cases; therefore it is essential to select a system that suites the case under investigation (CIRIA, 2001).

With regard to the strategic framework illustrated in Figure 9.1, strategic environmental assessment at national and regional levels should examine operational plans and action programmes. These will be concerned with the way economic, social and environmental considerations are integrated and applied to encourage sustainable efficiency and waste reduction. Embodied in this approach will be the assessment of effective land use and the balance between urban, semi-urban and rural development and the provision of sufficient infrastructure to meet laid down objectives.

Assessment systems are still in their early stage of development; however, attempts are underway to evaluate the effectiveness of decision-making so that lessons can be learned and incremental improvement can take place.

Environmental or thematic indicators

One way forward is to develop lists of relevant criteria that can be allocated weightings and scored to achieve assessments that can be aggregated to provide an overall assessment. For example, the following headings might be used:

1. Land use
2. Transport
3. Power generation
4. Energy consumption
5. Environmental pollution
6. Human capital
7. Natural capital
8. Commerce and business
9. Society and community
10. Ecosystem.

Developing this approach further, Figure 9.5 illustrates the main Theme Indicator Framework developed by the United Nations Commission on Sustainable Development. Division for sustainable Development (2001).

Each of the themes described above can be further developed into sub-themes and indicators. These provide the means to measure the effects of human activity on the environment as illustrated in Figure 9.6.

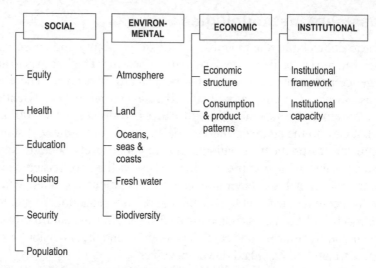

Figure 9.5   Outline theme indicator framework for assessment.
Source: UN Commission for Sustainable Development.

Atmosphere	INDICATORS
Climate change	Emissions of GHGs
Ozone layer depletion	Consumption of ozone depleting substances
Air Quality	Ambient concentration of air pollution in urban areas

Land	INDICATORS
Agriculture	Arable & permanent Crop Area Use of fertilisers Use of pesticides
Forests	Forest as a % of total land area Wood harvesting intensity
Desertification	Land affected by desertification
Urbanisation	Area of urban formal and informal settlements

Oceans, Seas and Coasts	INDICATORS
Coastal Zone	Algae concentration in coastal waters
Fisheries	Annual catch by major species

Fresh Water	INDICATORS
Water quantity	Annual withdrawal from ground and spring water as a % of the total available
Water quality	BOD in water bodies Concentration of faecal content in freshwater

Biodiversity	INDICATORS
Ecosystem	Area of selected key ecosystems Protected area as a % of total area
Species	Abundance of selected key species

Figure 9.6   Key environmental indicators.
Source: UN Commission for Sustainable Development.

240

Environmental impact assessment (energy)

Account must be taken of effects on the atmosphere, water, land, flora and forna and the built environment, plus the effects on humans. All options should be appraised and assessed as part of the decision-making process in the short, medium and long term. A thorough appraisal requires that all policy implications are considered in accordance with declared aims and objectives.

In the case of power generation this will include policies and regulatory activities that have an impact on the environment and may include pollution, waste, combination of fuels used and their relative consumption. The greatest environmental impact will arise from GHG emission identified as:

- Carbon dioxide (CO_2) produced from power generation and end use
- Methane leaks from natural gas distribution systems
- Sulphur hexafluoride leakage from electrical distribution grids.

The most significant GHGs covered by the Kyoto Protocol are shown in Table 9.2.

Table 9.2 Greenhouse gases

Greenhouse gas	Abbreviation	Source in gas and electricity industries
Carbon dioxide	CO_2	Generating stations, gas end use, compressor stations
Methane	CH_4	Pipeline leakage, generating stations, compressor stations
Nitrous oxide	N_2O	Generating stations, compressor stations
Hydro fluorocarbons	HFC	None from gas or electricity industries
Perfluorocarbons	PFC	None from gas or electricity industries
Sulphur dioxide	SF_6	Leakage from high voltage switchgear

Source: Ofgem.

The global warming potential of GHGs, as well as the quantity of emission must be taken into account, e.g. methane has 21 times the effect weight for weight than carbon dioxide. The principal source of sulphur dioxide is the burning of fossil fuels and there have been improvements in recent years due to a reduction in coal burning for domestic, industrial and power-generation purposes.

Table 9.3 shows the emission factors that allow a comparison to be made between the primary fuels that produce GHGs.

Road traffic produces large quantities of nitric oxide and nitrogen dioxide representing approximately 50% of all emission of these gases designated as NO_x. Invariably the largest concentrations are in urban areas

241

Table 9.3   Emission factors for the major greenhouse gases

	Natural gas	Oil	Coal	Nuclear	Renewables	Unit
CO_2	14 000	1900	2400	0	0	g/GJ
Assumed conversion efficiency	51	29	33	37	N/a	%
Emission factor	0.10	0.24	0.26	0	0	t C/MWh
Emission factor	0.36	0.86	0.96	0	0	CO_2/MWh

Source: DUKES 2000 Appendix B.

where traffic is at its heaviest. Table 9.4 shows the weight of NO_x and sulphur dioxide (SO_2) pollution from power generation in the UK between 2000 and 2001.

Table 9.4   Emission of No_x and SO_2 from energy emissions in the UK between 2000 and 2001

	2000		2001	
	NO_x	SO_2	NO_x	SO_2
Total emitted (kt)	365	821	379	743
g/kWh generated	1.00	2.25	1.04	2.04

Source: Digest of Environmental Statistics.

In addition to atmospheric pollution there is some contamination of inland and coastal waters caused by oil leakage from electricity substations and oil-filled underground cables.

Of principal concern is the amount of radioactive waste produced from nuclear power stations that represents a very high proportion of all radio-active material produced. Power generation produces different levels of radioactive waste. At the highest level is waste resulting from the repro-cessing of nuclear fuel and there is no means of disposal other than safe storage. Intermediate level waste includes reactor casings and components that similarly have to be securely stored. Low- level waste includes protec-tive clothing which can be disposed of in special facilities. Very low-level waste can be disposed of in landfill or it can be incinerated.

Other issues

There may be local environmental issues concerning noise created along the gas and electricity supply chain and these require dealing with on an individual basis. Power lines have also been cited as health hazards to nearby residents, but as yet evidence supporting this is inconclusive.

The power industry generates ash arising from combustion at power stations. Most of this is recycled by selling it to the construction industry to manufacture lightweight concrete building blocks.

There are landscape impacts that apply to traditional power stations and renewable power generation. In the past open cast coal mining was an issue that required reinstatement, but this is now less of a problem due to the reduction in burning coal. Apart from land pollution a major concern relates to the creation of blots on the landscape created by power stations, overhead power lines and unsightly animations caused by wind turbines.

Consideration needs to be given to the impact on wildlife and biodiversity. The impact of atmospheric pollution from power generation and end use is the major concern. However, the disposal of waste and land contamination should be regularly checked to establish the affect on ecosystems and early warning signs should be detected and action taken.

Assessment for building-type infrastructure

Assessment techniques are intended to demonstrate or compare effects of building-type infrastructure on the environment. This could be applied to different types of trade infrastructure (offices, factories, supermarkets, etc.), and social infrastructure (houses, churches, leisure facilities, etc.). There are several different systems that are designed for specific buildings and circumstances. Arguably the best known in the UK is a series of rating systems for different types of building known as the Building Research Establishment Environmental Assessment Method (BREEAM). The scheme was launched in 1990 and covers industrial buildings, offices, homes, supermarkets and does have the ability to be adapted to cover a wider range of buildings.

The objective of BREEAM is to mitigate the effect of buildings on the environment and to provide a measure indicating the extent to which environmental issues can be addressed. The point scoring system utilised also has the added advantage to facilitate comparison of the environmental performance of different buildings. The categories of approval associated with certification are pass, good, very good and excellent.

The 2004 version of BREEAM concerning office approvals allocates credit points under the following category headings:

- Management: overall management policy, commissioning, site management and procedural issues.
- Health and well-being: internal environment, sick building syndrome, lighting, workstations and noise levels.
- Energy: metering, insulation and CO_2 emissions.
- Transport: transport related CO_2 and location-related factors.
- Water: consumption, metering, leakages and waste prevention measures.
- Materials: presence of asbestos, use of recyclable materials.

- Land use: green or brown field sites.
- Ecology: ecological value conservation and enhancement of the site.
- Pollution: air and water pollution issues.

A checklist of potential scores is provided covering all 10 categories and the total points awarded for (i) 'core' and (ii) 'design and procurement' are kept separate.

The total score of (i) 'core' gives the probable environmental performance index score (EPI) as shown in Table 9.5.

Table 9.5   Determination of EPI

Checklist score	EPI
0–120	1
100–150	2
130–185	3
165–220	4
200–255	5
235–290	6
270–325	7
305–360	8
340–395	9
375+	10

The scores from (i) 'core' and (ii) 'design and procurement' are then aggregated to provide the BREEAM rating as shown in Table 9.6.

Table 9.6   Determination of BREEAM rating

Checklist score	BREEAM rating
235–405	PASS
385–550	GOOD
530–695	VERY GOOD
675+	EXCELLENT

BREEAM has the potential to be used by clients, planners, developers and agencies to specify the sustainable performance of buildings and to develop action plans, as well as monitoring and reporting performance at local and portfolio levels.

It is argued that BREEAM offers a range of benefits listed as follows:

- demonstrates compliance with environmental requirements from occupiers, planners, development agencies and developers;
- environmental improvement by supporting a wider corporate strategy or as a standalone contribution;

- occupational benefits by providing a better place for people to work and live;
- better marketing using ratings as a selling point to tenants and customers;
- the achievement of higher rental incomes and increased building efficiency;
- comparative analysis of performance between buildings; and
- greater capability to respond to client requirements.

To be awarded a BREEAM rating the assessment must be carried out by a properly trained and qualified BREEAM assessor. More information can be obtained from the BREEAM website referred to at the end of this chapter.

There are numerous rating systems that have been developed in North America, Europe, Asia and Australia, for example, in the USA systems include:

- Green building council's leadership in energy and environmental design (LEED)
- Building for environmental and economic sustainability (BEES)
- Energy efficient mortgage programme (EEM) aimed at residential homes.

Justification of sustainable construction

The credibility of sustainable construction rests with its ability to prove that the benefits to be derived have embedded value concerning the contribution to a better environment, increased performance and an adequate return on investment (ROI). Financial arguments tend to predominate, especially when they hold sway over whether a project is feasible or not. In the past, there has been considerable importance attached to the capital cost of projects, rather than life cycle and performance costs. This is specifically the case were construction projects are speculatively developed for immediate disposal on the market when complete. In these circumstances short-term considerations predominate, since the developer can walk away from the consequences of life cycle costs without penalty, once the project is sold. Property development in Hong Kong in the late 1980s represents a good example where gross development values (GDVs) were primarily driven by the cost of land and growth in the economy, which resulted in good ROIs within 7 years. Hence, redevelopment requiring demolition became financially feasible long before property had reached its design life expectation. This is an extreme example of financial imperatives overriding environmental and sustainable considerations.

More recently, the World has become aware of the finite nature of non-replaceable resources and damage inflicted on the environment. These are the beginnings of a realisation that sustainable infrastructure and buildings requires a more sophisticated approach, which to date has been mainly voluntary and is dependent upon a distinct change in attitude (Whittingham et al., 2003). There also needs to be a greater awareness and acceptance of the consequences of pursuing unsustainable growth. Hence, the need to communicate the importance of sustainability to the survival of the human race, and all other species and life that inhabits this planet. Linked to this must be an awareness campaign to demonstrate that sustainable solutions can provide significant tangible benefits without inflicting crushing economic burdens.

Case studies

Case study 9.1: The Sigma project

The Sigma Project was launched jointly by the British Standards Institution, Forum for the Future and Accountability in 1999 with the intention of exploring how organisations can learn from past experience with a view to taking a coherent approach to sustainability both at present and in the future. The project is funded by the British Government through the Department of Trade and Industry and is enhanced by a diversity of national and multi-national organisations that have set up stakeholder and steering groups. The idea is to enable organisations to become more sustainable by addressing the triple link effect of social, economic and environmental imperatives. The objectives of Sigma are therefore:

- To develop a management framework to help organisations to tackle sustainability. This involves practical guidelines and tools that empower organisations to adopt more sustainable solutions to doing business.
- To develop a management process that embeds sustainability in mainstream organisational policy, strategy, practice and procedures. The approach is to develop a tool that encourages more holistic and integrated thinking towards sustainability.
- To establish and nurture learning communities that further the aims and objectives of sustainable development with due recognition of stakeholder views.

The work of Sigma draws on internationally recognised best practice and works with the global reporting initiative (GRI) and the World Business Council for Sustainable Development (WBCSD). Expertise and views are also drawn from other international initiatives including natural step, accountability 1000 and ISO standards.

Recognition is given to the fact that all organisations need to develop their own strategy based on a holistic and integrated understanding of sustainability that takes a long view with suffice breadth and depth to cover most possibilities. Further, the approach taken should be sufficiently robust to account for change and developments in technology.

The Sigma project aims to unlock thinking that stands between understanding and action in many organisations by:

- integrating social, economic and environmental issues into strategic decision-making and opportunities;
- seeking competitive advantage through the sustainable approach;
- identifying and learning impacts and risks associate with activities;
- preventing, mitigating and managing risks and opportunities;
- identifying opportunities for continuing improvement;
- engaging stakeholders in decision-making processes; and
- encouraging the use of indicators to measure and compare performance.

At the core of the Sigma project are guidelines comprising a series of interlinking and supporting components, including principles, a management framework and tools.

The Sigma Principles are the ground rules of sustainability in that they underpin the concepts of accountability, capital enhancement and environmental sustainability. They also pose questions relating to key stakeholders and how they deal with social and economic capital.

The principles are drawn from a range of internationally accepted doctrines, conventions and declarations including:

- The UN Global Compact
- The International Chamber of Commerce Business Charter on Sustainable Development
- The Rio Declaration on Environment and Development
- The Global Sullivan Principles
- The Organisation for Economic Co-operation and Development (OECD): Guidelines for Multinational Enterprises
- UN and ILO conventions
- The Natural Step
- AA 1000.

The Sigma Principles establish the basis for an organisation's contribution to sustainability.

The Sigma Framework provides a process driven approach by means of a number of flexible implementation phases starting with a management phase that provides a baseline review that establishes organisational values, strategies and performance as they relate to sustainability. This is followed by the integration of sustainability and stakeholder engagement into core processes and decision-making and the development of an understanding of the relationships between organisational actions, impacts and actions and how they are managed.

The next phase is to formulate long-term strategies followed by the development tactical plans supported by adequate communication and training measures that support all levels of implementation. Control measures require development to ensure that actions, impacts and outcomes work in alignment with and support all levels of planning.

To meet the needs of internal and external stakeholders and to translate feedback into appropriate change progress reporting of review mechanisms requires to be developed.

The Sigma Toolkit supports practical application of the SIGMA Framework by providing:

- A business case for sustainability
- A Sigma Management Framework Benchmark Questionnaire
- The Sigma Sustainability Scorecard
- A stakeholder engagement process based on AA1000
- An approach to environmental accounting
- The GRI approach to performance measurement and sustainability reporting
- A tool that illustrates the contribution of existing systems
- Standards and approaches to SIGMA guidelines
- A summary of supporting standards, guidelines and tools.

The concept on which Sigma is based relies on companies, NGOs, research organisations, local authorities, higher education institutions and other stakeholders providing feedback. As more organisations become involved in Sigma it is anticipated the principles and guidelines will evolve and mature. More information can be found about this project on http://www.projectssigma.com

Author's commentary

The Sigma project is a good example of the sort of initiative that is required to support the awareness, understanding and implementation of the so-called 'Triple Bottom Line'. Sigma is now offering a one-day Master Class and a four-day Practitioner's Course. More information can be obtained online at: www.bsi-global.com/seminars

Case study 9.2: Renewable energy for local communities in the USA

Despite the decision by the government of the United States not to ratify the Kyoto Protocol it is significant that a bottom-up initiative is gaining ground in Minnesota and to a lesser extent in Wisconsin, Iowa Illinois and Massachusetts. There are also initiatives in New York and the Western United States. The US government has been slow in supporting the use of wind power, whereas by contrast the European Union has adopted policies that support energy generation by wind power resulting in widespread commissioning of large wind farms (Bolinger, 2001) with a target of 10% of total demand coming from renewable resources.

In areas of the US that have good wind profiles farmers have always been interested in wind power, but tax-based incentives introduced by the US government have been aimed at corporate entities with large tax credits. More recently, with assistance from state policies, particularly in the upper Midwest interest has been created in wind power generation from local farmers, schools, towns and individual investors. The main issues to be considered relate to the size of project, the ability to sell excess power to the national grid, the type of ownership and interconnection.

Discounting large wind farms, this study is concerned with small-scale wind power development of 2 MW or less, normally involving one or two turbines. In Minnesota favourable conditions for investment have been created by the availability of suitable wind turbines, together with grants and tax breaks. There have also been offset agreements, for example, in exchange for supporting wind power development, a power generator has been granted concessions about the storage of nuclear waste. It is possible that up to 50% of the cost can be covered by grants, but this still leaves typically in the region of US$ 1 million to be found by investors.

The ideal situation is for these wind projects to be locally owned by community shareholders who have the right to participate in the benefits of the scheme. This also has the advantage of overcoming local objections on the grounds of the visual impact and noise of turbines. The alternative is for projects to be funded through normal commercial channels. There are also innovative schemes, whereby a wealthy individual investor puts up the cash and takes advantage of the benefits and tax credits for the first 10 years, then transfers ownership to local investors.

In Minnesota the aim is to provide at least 460 MW of community wind power and up to January 2004 a total of 132 MW had been built with another 68 MW likely to come on line.

In Wisconsin wind resources are not as plentiful as Minnesota and there are a limited number of projects. Typically, projects are small scale, for example, there are two 1.65 MW turbines in Eden town. Likewise in Illinois the Illinois Clan Energy Community Foundation has supported school projects, e.g. a grant of US$ 35 000 was given to Valley District School to undertake a feasibility study. This proved to

249

be positive and a construction grant was given amounting to US$ 331 678 to build a 750 kW turbine at the school. The grant represented in the region of 35% building costs.

Author's commentary

This case study demonstrates that it is feasible for local communities, with adequate federal and state government support to take responsibility for producing their own power from abundant renewable resources. The key requirement is the amount of wind power available throughout the annual cycle and the ability to account for the exchange of power to the grid. Given the somewhat intrusive nature of wind turbines local ownership is very important to overcome 'the not in my backyard syndrome'.

Summary

This chapter demonstrates the need for a global approach to sustainability as proposed by the Earth Summits in Rio, Kyoto and Johannesburg and it is regrettable that to date some countries, including the USA, have refused to cooperate. The European Community has taken a strong lead on addressing sustainable issues and the threats posed by global warming. Current actions proposed in Europe provide an excellent example for the remainder of the global community to follow.

The need for the development and adoption of sustainable strategy that is both holistic and integrated incorporating the provision of infrastructure and the wider built environment has been articulated. The strategy proposed incorporates a layered approach, whereby regions and local communities are involved using a cause and effect system that takes a broad view across the triple-link concept in a transparent and accountable manner.

The need to address issues associated with global warming and climate change has been explained and an account has been provided of the steps that have taken place or are being planned in Europe. GHGs have been identified, together with their potential to cause global warming and the case for moving towards a low-carbon economy has been made. Issues concerning transport, energy, water systems and waste disposal have been highlighted and explained.

The importance of proper environmental impact assessment and risk appraisal has been stressed and examples have been provided using the United Nations Sustainable development Indicators and the BREEAM Assessment system developed in the UK for buildings.

The chapter concludes with a justification for sustainable infrastructure supported by two case studies. Reference should also be made to Chapter 11 that looks at future technological and conceptual developments in the provision of infrastructure.

References

Bolinger, M. (2001), '*Community wind power ownership schemes in Europe and their relevance to the United States*', LBNL-48357. Lawrence Berkeley National Laboratory, Berkeley, CA.

Brundtland, G.H. (1987), '*Our common future, world commission on the environment and development*', Oxford Paperbacks, Oxford, UK.

Building Research Establishment (2004), '*BREEAM Offices 2004 Manual*', online at www.bre.co.uk.

Construction Industry Council (CIC) (2003a), '*Integrated transport and land use planning*', Thomas Telford, London.

Construction Industry Council (2003b), '*How buildings add value for clients*', Thomas Telford, London.

Construction Industry Council (CIC) (2004), '*Constructing for sustainability: a basic guide for clients and their professional advisors*', CIC, London.

Construction Research and Information Association (2001), '*Sustainable construction indicators*', C563, London.

Construction Research and Information Association/Construction Industry Council (CIRIA) (2003), '*Brownfields – building on previously developed land: a briefing guide for construction clients*', ISBN 1 89867 1214.

Division for Sustainable development (2001), '*National information indicators*', United Nations, available at www.un.org.

Ekins, P., Hillman, M. and Hutchinson, R. (1992), '*Wealth beyond measure: an atlas of new economics*', Gaia Books, London.

Farrall, S. (1999), '*Financing transport infrastructure: policies and practice in Western Europe*', MacMillan, London.

Forum for the Future (1999), '*The Sigma project*', online at www.project-sigma.com.

Griffiths, P.L.J., Smith, R.A. and Kersey, J. (2003), '*Resource flow analysis: measuring sustainability in construction*', Proceedings of the Institution of Civil Engineers 156, September Issue ESI, pp. 142–152.

Institution of Civil Engineers (2004), '*Safeguarding our future*', ICE, London.

Kibert, C. (1994), '*Establishing principles and a model for sustainable construction*', Proceedings of the 1st Conference on Sustainable Construction, Tampa, Florida, USA, November, pp. 3–12.

Ofgem (2003), '*Ofgem's approach to environmental impact assessments*', online at www.ofgem.org.uk

Parkin, S. (2000), '*Sustainable development: the concept and the practical challenge*', Civil Engineering 2000, 138(Special issue), 3–8.

Parkin, S., Sommer, F. and Uren, S. (2003), '*Sustainable development: understanding the concept and practical challenge*', Proceedings of the Institution of Civil Engineers 156, March Issue ESI, pp. 19–26.

Pitts, A. (2004), '*Planning and design strategies for sustainability and profit*', Architectural Press, Oxford, UK.

Serageldin, I. and Stee, A. (1994), '*Expanding capital stock in making development sustainable: from concepts to action*', ESD Occasional Paper Series No. 2, World Bank, Washington, D.C.

Sustainable Task Group (2004), '*Making the most of our built environment*', online at www.dti.gov.uk.

Whittingham, J., Griffiths, C.S. and Richardson, J. (2003), '*Sustainability accounting in the construction industry*', Proceedings of the Institution of Civil Engineers 156, March Issue ESI, pp. 13–15.

Chapter 10

Facility management of infrastructure

Introduction

The theory and practice of facility management came into prominence in the mid-1980s as a function that recognised the importance of efficiently planning and managing the use of built infrastructure facilities throughout their life cycle from inception to disposal. New concepts were introduced focussing on operational standards such as space planning, integrated building services, energy efficiency and planned maintenance. Importance has also been attached to the impact of changes in technology and the need to adapt facilities to meet changing user requirements.

The recognition of facility management has raised the profile of the debate concerning capital investment in infrastructure and the subsequent operational and maintenance costs required throughout the life cycle (Park, 2004). As a consequence, design quality has come under increased scrutiny, especially with regard to functional performance and operational efficiency. Facility management requires extensive investigation into all operational factors and it is therefore necessary to use tools and techniques such as performance evaluation, financial appraisal, risk assessment and sensitivity analysis, together with the evaluation of options concerning additional capital expenditure intended to reduce operational and maintenance costs. Typical operating costs can be up to 10 times or more compared to capital expenditure and this makes it essential that adequate consideration is given to the selection of the most efficient designs intended to fulfil client needs.

The role of the facility manager involves a far wider range of activities compared to just maintenance and repair alone. The facility manager is, in addition, expected to understand all aspects concerned with facility acquisition, performance and space utilisation and to communicate and coordinate with the design team in the provision of new and extended facilities to accommodate changing needs. Facility management can be provided 'in-house' or it may be outsourced to a specialist organisation.

The introduction of new forms of procurement involving various levels of public–private partnerships (PPP), where revenue is derived from rent, user fees or toll charges to cover prior capital expenditure, has further enhanced the importance of facility management. A comprehensive awareness and understanding of the operational implications of different types of infrastructure design such as user needs, performance in use and life cycle costs (LCC) are crucial to judgement of the feasibility, viability and risks associated with long-term PPP projects. It is therefore important that the design of infrastructure facilities takes into account user functions, performance benchmarks and the costs associated with maintaining the facility in compliance with service specifications. In the case of BOT or DBFO projects, the contractor owning the concession to operate a project will need to gear up for facility management by covering all necessary operational requirements. Hence the quality and efficiency of facility management will have a direct impact on profitability and success.

This chapter discusses the theory and principles of facility management and then suggests how it should be applied to the operation of infrastructure facilities. It starts with the operational context defining facilities management and its scope, the case for facilities management and the professional skills required to fulfil such a vital function. Attention is given to key operational factors such as the impact on design practice and the need for performance indicators. A framework for facilities management highlighting the key factors to be considered including statutes and legal requirements is discussed. Coverage is given to life cycle appraisal techniques and methods necessary to assist valid judgements about design quality and performance, discounted future costs, revenues and risk associated with long-term investments in infrastructure projects. The chapter concludes with the practical steps to be taken in operating infrastructure facilities such as understanding the role of the client, packaging and outsourcing contracts, service level and contractual arrangements, facilities condition register and the need for financial management and cost control systems.

Operational context

Definition and scope

The main operational tasks concerned with facilities management (FM) extend from completion of the construction phase and handover of infrastructure projects to their final operational function and disposal. In addition, it is now widely recognised that life cycle operation, maintenance and repair of infrastructure facilities must be considered from inception and throughout the design period. FM encompasses multiple disciplines that require careful co-ordination to achieve functionality according to specified benchmarks and performance targets. This involves the client and users as well as those responsible for operation, upkeep and repair. The role of the facility manager is to integrate people, place, process and technology. Core competencies are required concerning appropriate technology, leadership, finance, planning and project management, communication, operations, maintenance and repairs (Spedding, 1994).

There is a growing realisation that conceptual, scheme and detailed design decisions have a fundamental impact on operational, maintenance and repair costs. Normally this will require trading off increased capital expenditure against lower operating and maintenance costs. The application of discounted cash flow (DCF) and net present value (NPV) will considerably assist design decisions; however, designers and clients should always strive for efficient aesthetically pleasing designs that maintain functional performance standards at no, or least, extra cost. The experience of facility managers, together with their knowledge and expertise will assist the design team in developing designs where total LCC can be accurately predicted and where performance benchmarks can be reliably determined.

Facility management may therefore be defined within the context of infrastructure projects as follows:

'The practice of co-ordinating the physical working environment of an infrastructure facility with people and processes in accordance with the objectives of the client organisation by the integration of business, design, engineering and behavioural sciences'

The components of facility management consist of an organisational and managerial core that interacts with major components concerned with strategy, finance, operational control and contractual arrangements. These should be viewed within a life cycle environment as shown in Figure 10.1.

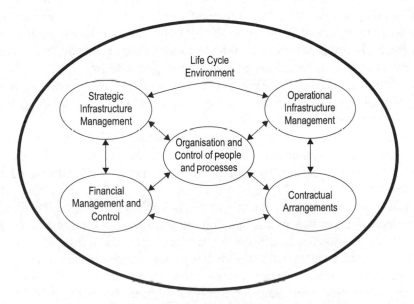

Figure 10.1   Components of infrastructure facility management.

255

Facility management will vary in nature and extent depending upon the type of infrastructure. The most dramatic difference occurs between linear infrastructure projects, e.g. roads and permanent way rail track and infrastructure confined to one or more sites, e.g. power stations and dams. Invariably infrastructure consists of a combination of both linear and site-based components where bridges, tunnels and ancillary buildings are included. Another major distinction concerns projects that provide enclosure and internal environments for human work, care and living or a combination of all three, as in the case of hospitals and residential nursing homes. In such cases services to maintain the internal environment will be a major factor alongside space utilisation, security and safety and the need to incorporate medical and scientific equipment and facilities. Often functions will be multiple and complex requiring a thorough knowledge of cause and effect between different processes and activities.

Linear components are either exposed to the elements or buried underground. Normally function will be singular and subjected to constant use, e.g. expressways and power lines. Key to the satisfactory and safe function of such facilities will be regular inspections and planned maintenance and replacement schedules, which should exploit periods of low demand where limited service provision will be adequate. This implies careful planning and speed of operation.

The case for facilities management

Recent research has proved that life cycle operation, maintenance and repair costs can exceed the initial outlay of capital expenditure by up to a multiplying factor of 10 or more. This is compounded by the need to conserve energy and to make the best use of non-replaceable natural resources, including potential to recycle materials at the end of the life cycle. Consideration should be given to the impact of current trends such as an increasingly litigious society and the threats posed by crime and terrorism. There is also an increased need for security and safe travelling, working and living environments. As a consequence clients have become more sophisticated and demanding in their approach to commissioning and procuring infrastructure.

Risk associated with investment in infrastructure is a major consideration and this can be mitigated by meticulous evaluation of functional performance and efficiency within the systemic diagram illustrated in Figure 10.1.

Given a choice of design scenarios it is possible to examine and evaluate options using sophisticated simulation models incorporating visualisation and financial modules to determine the best potential solution. The use of value analysis, cost benefit and sensitivity analysis may present the potential to make further iterative gains (Barratt and Baldry, 2003). Savings made during the design stage have potentially more benefit than trying to achieve gains during the construction stage. Hence it is vitally important that the operational knowledge of an experienced facility manager is sought as early as the development of the conceptual design. In this manner, it is more likely that the infrastructure proposed will fulfil and, in some cases, exceed functional performance specifications. Where changes and developments in technology are likely to occur these should be anticipated in order to allow for infrastructure provision to be upgraded with minimum cost and disruption. Such provision will normally rely on the foresight and vision of the design team.

Professional skills of the facility manager

It is essential that a facility manager is in possession of a hybrid of technical and managerial skills, as well as sufficient knowledge of the various technologies involved and how they interact and integrate to achieve in-use function and performance. Nevertheless, the facility manager's performance will always be judged on the quality outcome of managerial capabilities rather than technical competence in one or more disciplines. Rather, the facility manager's technical knowledge should be sufficient to enable greater appreciation of problems, or to signal the need to call in specialists to give advice. A major factor contributing to the success of a facilities manager lies in the ability to achieve the correct balance of managerial and technical knowledge. This will vary from project to project and it is therefore necessary for an appraisal to be made of the managerial and technical requirements of each project from the outset. Where there is perceived to be a shortfall, steps should be taken to acquire the necessary information and skills. Vision and foresight to anticipate developments in technology, practice and client needs will assist the facility manager to perform a strategic role whereby strong interaction takes place with the design and construction teams during the pre-occupation phase. Post-occupation facility management is a core management activity that requires exceptional skills concerning communication and co-ordination integrated with strong leadership.

The responsibilities of the facility manager will depend to some extent on the procurement method adopted for individual projects, but in its fullest mode FM is about the effective and efficient delivery of all that concerns the operation and function of built infrastructure facilities. This requires organisation and motivation from the FM service provider team that accords to purchasing and contractual arrangements. Good contractual relationships are key to successful FM and much will depend on personal and interpersonal skills. It is therefore important that the facility manager has the ability to

- identify client and user needs;
- develop facilities policy that reflects needs;
- communicate policy and its implementation;
- develop and implement procurement strategies;
- determine and negotiate service level agreements;

- select and implement purchasing and contractual arrangements;
- develop teamwork through partnership;
- systematically monitor and appraise performance; and
- take actions to achieve continuous improvement in the quality of service.

In conclusion, the facilities manager's role is a core activity that contributes to the success of built infrastructure facilities according to their function and the level of service provided to clients and users. The approach adopted should be open and transparent where issues are surfaced and addressed in a positive mood, while giving recognition for effort and contribution by team members.

Interaction with design in practice

The facility manager should ideally be involved throughout the design process. The consequences of design decisions are recognised as being fundamental to LCC and functional performance. Therefore, the design team require the benefit of specialist knowledge and performance data that provides the foundation for the evaluation and appraisal of construction solutions involving a multiplicity of materials, components, fabrications construction processes and sequences (Alexander, 1996). Alternative designs, materials and components should be examined to determine cost benefit and value added. The aim should be to optimise and select the best design solutions according to functional needs and performance specification, where necessary, testing, simulation and prototyping should be used to accredit performance, especially where innovation and untried design solutions are being considered.

Performance indicators

It is necessary to measure the performance of construction projects to facilitate continuous improvement by means of comparison with other similar projects during construction and through the operational life of the completed infrastructure facility.

This normally includes external benchmarking where comparison is made with other infrastructure projects to determine

- assessment of client's performance;
- a framework for performance measurement that includes cost and time predictability, client satisfaction, number of defects, safety record;
- other measures that might include variations to the project and the final cost compared to the original estimate; and
- project-specific measures that might include success with the achievement of cost reduction through innovation, sustainability, lower energy consumption, fitness for purpose, user satisfaction and lower maintenance and operational costs.

The effectiveness of infrastructure projects can initially be measured by determining value against quality and cost, together with speed of construction. It is also essential to extend measurement to whole life cycle performance. Initially, costs will be predicted according to the best available performance data and these will form the basis of operational benchmarks and targets. The achievement, or otherwise, of these benchmarks must be monitored to form the basis for reliable future data that will determine the tasks required to achieve continuing improvements to operational performance and costs.

Design quality indicators

The debate on design quality has been enhanced by the development of a Design Quality Indicator (DQI) by the Construction Industry Council (CIC) in the United Kingdom. It is a unique method of evaluating the quality of built facilities using structured questionnaires to elicit data from a wide variety of actors concerned with the production of the Built Environment. The intention is to steer and influence the design process, align the design requirements of project participants, assess the quality of the project at all stages and to help achieve greater value from design (CIC, 2002a,b). There are three headline DQI indicators as shown in Figure 10.2.

Functionality	Build Quality	Impact
Use	Performance	Form & Materials
Access	Engineering Systems	Internal Environment
Space	Construction	Urban & Social Integration
		Character & Innovation

Figure 10.2   Three DQIs.

At the inception stage the DQI sets the framework for the consideration of product quality by informing the client and the users about choice. DQI then assists the improvement of design quality by evaluating different scheme proposals. At the construction stage the DQI tool facilitates integration between team members and when the built facility is complete and occupied it can be applied to evaluate the perception of users and others to provide a better understanding to inform future projects. Hence the main purpose of the DQI is to act both as a comparator and an indicator, thus enabling comparison between the relative performance of projects. Figure 10.3 illustrates the DQI visualisation tool, which is essentially a radar chart that plots each of the DQIs on a scale from basic to excellent. In this manner, projects can be compared at various stages by overlaying the profile of results. Relative strengths and weaknesses can be determined and scope for improvement can be identified.

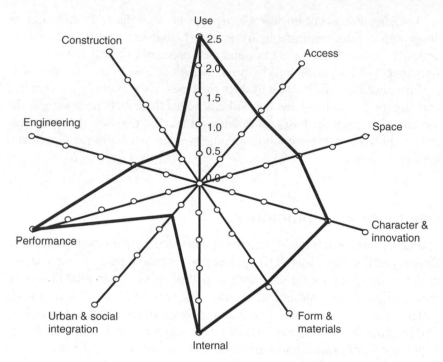

Figure 10.3   The DQI visualisation (0.0 basic – 2.5 excellent).

Another approach is to develop a set of key performance indicators intended to provide an ongoing check on how a project is performing in the key stages of its life cycle. Such reviews and checks may be carried out in-house, across disciplines, organisations and clients. Fundamental to these approaches is a common measurement scale that can determine norms and benchmarks. In this manner areas of weakness can be determined in a timely manner and the appropriate action taken to effect improvement. A major advantage associated with radar charts is that they provide good visualisation and rapid understanding.

Facility management framework

Facilities management can be considered at various levels commencing with a single project where the parameters limit consideration to the goals of the project and their achievement. Beyond this the facility management of multiple projects leads to local, regional and national policies and the strategies necessary to achieve strategic aims and objectives. Large public and private owners of infrastructure require a systematic understanding of infrastructure holdings and it is essential for them to pursue an explicit management strategy that seeks out economies of scale, standardisation and unified practice that enables the application of systems to achieve efficiency, waste reduction and quality services (Langston and Kristensen, 2002).

260

Portfolio holdings of infrastructure

A portfolio holding of infrastructure may comprise a regional expressway network, an area health care network comprising hospitals, clinics and primary care centres, power generation plants and a regional distribution system. In all cases, it is important to understand how the various parts of the infrastructure portfolio contribute to the overall strategic aims and the achievement of well-defined operational objectives. Within this context, the facilities will need to be managed in unison requiring policies and strategies geared to reaching benchmarks defined to establish the achievement of quality services.

Strategy for the provision of facility management

A vital prerequisite is that there must be recognition by clients, whether public or private, of the important role that efficient and well-maintained infrastructure plays in the achievement of strategic objectives. There must also be an acknowledgement of the importance of a coherent strategic plan with well-defined aims and objectives.

Such a plan should be developed using a comprehensive data and information framework comprising the following:

- Database defining existing provision, including a comprehensive profile of infrastructure, levels of service, expected remaining life, energy efficiency, LCC, sustainability.
- Demographic data and forecasts, economic forecasts, projected traffic flows.
- National, regional and local development plans including land use, urban development and rural protection.
- Government policy relating to health, education, transport, power generation and consumption, telecommunications and the provision of services, e.g. water treatment, sewage treatment, waste disposal and the control of atmospheric pollution. Attention should also be given to public safety and security.
- Other published and recorded data as conceived to be appropriate.

Using the above data a profile of existing and future demand for infrastructure services can be established that show areas of growth, stability and decline. Account can also be taken of developments in technology and practice that may impact on the sort of infrastructure that needs to be provided in future.

Once a basis for demand has been established feasible targets can be considered within identified time frames. Much will depend on the ability to generate sufficient capital using income and borrowing. Where needs are proved to be in the national interest, the means to leverage public and private funding can be established (see Chapter 7). Once the demand case has been accepted socially, economically and politically, then strategic proposals to

261

fulfil these demands can be developed. Alternatives will be considered, analysed and optimised to best achieve identified goals.

From the strategy will emerge project proposals that will realise stated aims and objectives. These proposals will be further considered in relation to existing infrastructure provision and operation. Once a project is initiated the facility manager should be engaged to work closely with the design team to ensure that all aspects concerning the efficient operation and function of new and existing provision are harmonised with a view to achieving continuous improvement.

In setting a facility management strategy, recognition must be given to the need for continuous monitoring and periodic review. There must also be an awareness of the changing needs of clients and users that require alterations to infrastructure provision and the introduction of new practices and procedures. Therefore, the prime objective should be to operate infrastructure in the most economic and efficient manner to support activities that contribute to the achievement of services at the level of quality required.

Procurement of facility management services

As described in Chapter 6 the procurement of facility management services may take a variety of forms. In the event of design and construction contracts that hand over completed infrastructure facilities in totality or in phases, it will be necessary to have a facility manager in place to implement operations. Under these circumstances, there are advantages in appointing a facility manager as a consultant to advise the design team and to work with the contractor as the work progresses. By using this approach the design solution will benefit from operational experience and the realisation of the project will be checked at various stages of completion to ensure that the finished product will actually work as intended. The facility manager will also play a key role in commissioning and handing over the project. Post occupation, the facilities manager's role will fundamentally change to procuring services and managing the efficient operation and the achievement of performance benchmarks through monitoring and control procedures (Atkin and Brooks, 2000).

Where projects involve 'design, build, finance and operate' (DBFO) type contracts, facility management expertise associated with operation and function should become a fundamental part of the design and construction process commencing at the concept stage.

The facility manager will play a vital role in assessing additional demands or changes requiring action to provide new or extended facilities in areas of expansion, while areas of decay may require downgrading or disposal.

Where works are relatively small these may be dealt with in-house by a small team employed directly by the facility owner, but in the case of larger works professional assistance should be sought to provide a design solution and to procure the completion of the necessary works. Figure 10.4 shows the traditional process to procure additional facilities.

In liaison with the client, the facilities manager will play a key role in the determination of need and the preparation of a brief. Once the design team has been appointed alongside a project manager, the facilities manager should participate in the development of the design brief. A successful brief will help to ensure that all client requirements will be fulfilled. Where appropriate new development will be undertaken on land already owned by the client, but in the event that a new location is required, or the existing site is not large enough, then new land may need to be purchased and if necessary planning and statutory permissions must be acquired.

The method of procuring infrastructure works as shown in Figure 10.4 assumes the appointment of a design team and a project manager before the selection of contractors. In the event that design and build and fee contracting

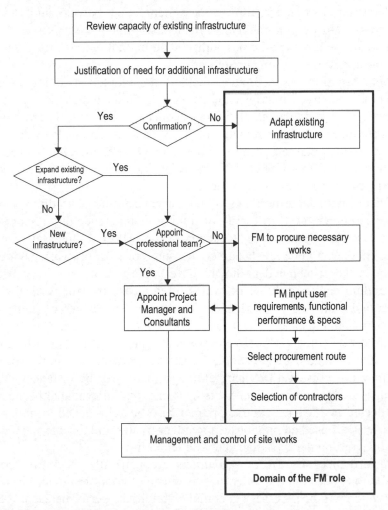

Figure 10.4   Traditional procurement of additional infrastructure.

procurement options are used, the process will require modification, but this will not change the requirement for there to be close collaboration and communication between the facility manager and the design and construct teams. Furthermore, where a PPP is involved, possibly using DBFO, then the whole process from inception to final decommissioning should be fully integrated within the control of one organisation (Akintoye et al., 2003). Details of the alternative methods of procurement are covered in Chapter 6.

In the event that the existing infrastructure facilities require alterations, the scale of the alteration and the potential disruption to users and processes will be key considerations. Often construction works are carried out in less than ideal conditions with restrictions on working space and hours, especially where existing services have to be maintained albeit at a degraded level. As a rule, design and build solutions are more risky and the traditional approach may under certain conditions, offer a better option provided that the design and working methods are properly thought through. The separation of construction work by temporary barriers and cleverly phased working programmes can mitigate disruption to both users and contractors and will reduce the possibility of additional claims brought about by non-productive working conditions.

Where traditional procurement methods are adopted involving competitive bids, selective tendering should be used to reduce the risk of cost and time overruns and to keep quality problems to a minimum. The vetting of suitable contractors is extremely important and should involve asking the right questions and taking up reliable references. Some clients prefer to maintain lists of approved contractors from which tenderers can be selected to bid based on a single-stage process.

Clients with large holdings of infrastructure may have direct labour organisations engaged to carry out routine maintenance and repairs; the alternative may be to employ term contractors who will undertake works, as required, using previously agreed measured rates or daywork payments. Recently, direct labour departments have fallen out of favour because of the random nature of the work, which often means that the work force is either under-utilised or overloaded. This increases the risk of inefficient and costly maintenance and repairs. The tendency has therefore been to favour term contracts whereby the contractor has the problem of best utilising labour. Where a term contractor has several contracts in the locality this may not present a problem, although there may be instances where labour is tied up on other contracts or it may not be available because of unexpected need. Term contractors may also be approached to undertake minor works based on previously agreed term rates and this has the advantage of avoiding delays associated with tendering.

Performance specification contracts are generally used for service installations because they offer the advantage of good design and installation, which will be necessary to comply with in-use performance as a contract condition. Using this approach, the facility manager will prepare a definitive performance specification that lays down air changes, power output, flow rates, temperatures, wear conditions, etc. These conditions must be fulfilled to successfully complete installation and formally comply with all the contractual requirements.

Statutes and legal requirements

Facility management involves a multiplex of legal requirements ranging from the acquisition and disposal of assets, legal and statutory conformity, employment legislation and requirements concerned with safety, health and security. This section provides an overview of the requirements, which will require further expansion according to the references given at the end of this chapter.

Urban and rural planning regulations

Central government will usually have a national master plan that determines policy for development. A key part of national policy will be the provision of necessary physical infrastructure to support economic development and growth. The main infrastructure elements contained in the master plan will be transport, seaports, airports, energy generation, water, waste disposal and recycling, telecommunications, healthcare, education and other public services. Other plans consisting of regional/state, county and local plans will provide more details regarding, for example, location, routes and provision for expansion. It is important that the respective responsibilities of national and local government are clearly defined. Where proposed infrastructure development is large and vital to the nation, it is likely that major development decisions will be taken by central government. In most democracies it is normal for prior consultation to take place with the public before a final decision is made. Therefore it may not be necessary for infrastructure development to go through the same planning approval procedures as in the case of industrial, commercial and residential development as set out below.

Development may be defined as undertaking construction, mining or other operations on, over or under land, or making a material change of use to existing structures, buildings or land. Given that there may be exemptions all activities of this type will require some form of planning permission. Procedures will vary according to national regulations that are set in place to ensure conformance with national, regional and local policies and plans. Normally land use will be zoned according to preferred use which may typically be classified as industrial, light industrial, residential, recreation, public open space, areas of natural beauty and agricultural. There will also be areas designated as of historic significance that will be granted heritage or conservation status. Planning applications that do not conform with designated land use will normally be subjected to increased scrutiny and a strong case will need to be made to counter rejection.

In most countries, smaller development will be dealt with by local authorities; however larger developments with area or regional significance may be determined at county or state level. Massive scenes affecting the national interest will invariably be finally approved by central government. Where applications are unsuccessful there should be in place an appeals system to ensure that appellants are dealt with fairly.

Enforcement of planning regulations is a key issue and it is generally accepted that the interests of the nation, businesses, social groupings and individuals are well served by the strict application of controls. Measures to stop illegal development and breaches in planning conditions will include the serving of enforcement notices. These must be complied within a stated time period, although there may be a right to appeal. Wilful non-compliance may attract significant fines and ultimately the imprisonment of those responsible.

The facility manager should be aware of the nature and requirements of permitted development under development control legislation. Normally, the acquisition of appropriate permission to extend existing facilities or to develop new sites requires set application procedures leading to determination. This process takes time and therefore early applications are advisable, since it may be extremely risky to acquire land or start development before formal permission has been given. Once planning permission has been granted, the facility manager should ensure that development takes place within the confines of the permission and where changes require further approvals, these should be dealt with in an expeditious manner.

Acquisition, ownership and disposal of land and property

It is usual for land or property to be acquired as freehold or leasehold. Freehold means for practical purposes absolute ownership in perpetuity, although ultimately all land is owned by the nation. Freeholds may also have covenants in place that place restriction on use. Leasehold means the right to use land or property for a period of time e.g. 25 to 99 years. There will always be a superior to the leasehold, whether it is a freeholder or a superior leaseholder. Leases normally contain covenants that place restrictions on the leaseholder, including permission to make alterations or changes of use. There may also be service charges imposed by the superior. At the end of the lease, the leaseholder must hand back the land or property to the superior leaseholder or freeholder.

Most countries operate a compulsory system of land and property registration whereby the title to land and property is legally defined and recorded. In this manner, disputes concerning ownership are largely avoided and governments are provided with an excellent means whereby capital gains taxation can be imposed on sale and acquisition.

The sale and acquisition of land or property invariably involves large sums of money and great care needs to be taken to ensure that transactions are properly carried out and recorded. Conveyancing should normally be placed in the hands of a lawyer, although this is not mandatory. The term 'let the buyer beware' is very appropriate and searches will be necessary to establish ownership, covenants, rights of way, nearby future development, subsidence, flooding, services, presence of toxic waste and previous land use. The condition of infrastructure, structures, apparatus and equipment should all be inspected and reported on. The outcome of these reports may mean that conditions are placed on contracts, which must be complied with before contract completion.

When selling it is normal to place land and property in the hands of agents, who will charge a commission once the sale has been successfully completed. The alternative is to use an auction, where the highest bidder is legally bound to purchase at the knockdown price. Clearly, bidders attending the auction should have previously carried out the necessary searches, surveys and reports. An alternative is a system of sealed bids where the highest open bid determines the legal buyer.

Where the acquisition of property or land is in the national interest, it may be subject to compulsory purchase where freeholders have no choice but to move out of property or land. Compensation will be determined by a district auditor or other authority who will be required to determine a fair market price for acquisition.

The facility manager may be asked by the client to arrange for the acquisition or disposal of land or property. It is therefore essential that the facility manager is aware of all procedures and legal requirements and be in possession of sufficient knowledge to liaise and negotiate with agents, lawyers and experts.

Health, safety, security and welfare

The facility manager is required to take a broad overall responsibility for health, security, safety and welfare that involves, employees, users and the general public at large. This involves considerable knowledge of legislation as it applies at all stages from the inception of infrastructure facilities to final disposal.

At the design stage there will be the need to create designs for safe occupation and operation as well as creating internal environments that are healthy and support a sense of well being from occupants. Designs will need to accommodate the safety of users, with the necessary back-up and emergency systems when things go wrong. Security to protect users and occupiers from fire, smoke and intrusion, as well as proper means of escape are other crucial requirements. Designs will also need to accommodate safe working conditions for site workers during construction by allowing working practices that conform to construction health and safety regulations. The requirements imposed by applicable construction and development regulations should be fully complied with, and checks should be in place throughout the design stage. Input from an experienced facility manager to the design team is highly desirable through the scheme and detailed design stages.

During the construction stage the prime responsibility for site safety will be with the contractors; however, the facility manager will become centrally involved where land or property is being altered or extended while the site is occupied by clients and their employees. The construction workspace should be defined and confined behind safety barriers, hoardings and gantries that provide safe access ways. Noise, dust and other forms of pollution should be strictly controlled and where dangerous construction operations are essential, this should be done when facilities are at low levels of activity or are shut down altogether. During the commissioning stage, it will be the role of the facility manager to work in close liaison with the commissioning team to ensure that systems operate as intended and that emergency equipment and procedures work effectively.

Once the facility is in operation, legislation will come into force regarding property, health, security, safety and health of occupants and users. There will also be the need to conform with employment legislation, data protection, equal opportunities, racism, civil rights and grievances. Safety systems should be properly maintained and inspections and procedures, including risk assessments must be regularly carried out. The facility manager will carry considerable responsibility throughout the operation of infrastructure and will, in many countries, be open to criminal prosecution in the event of negligence and incompetence. The updating of the infrastructure safety log through the life cycle will be the responsibility of the facility manager and this will play an important part in the final decommissioning, and where applicable, demolition and removal of the facility.

Life cycle appraisal (LCA)

There is a growing realisation associated with the importance of whole LCC and functional performance of infrastructure. This has been brought about by sustainability arguments and various research findings that have shown LCC to be up to 10 times or more than the initial capital cost of infrastructure. Further, the increasing popularity of PPP and the use of DBFO have encouraged the use of value analysis to examine the relationship between capital expenditure and running costs. This type of appraisal relies on the accuracy of predicting future operational conditions and costs, given an acceptable level of uncertainty. To assist decision making there are various risk analysis techniques, which can be used to identify the principal areas of uncertainty and their likely impact. Whole life cycle appraisal is about evaluating the feasibility of spending additional sums at the outset with a view to reducing expenditure in the future (Nutt and McLennan, 2000). This evaluation should take into account other benefits that may be derived from increased capital expenditure such as aesthetic appearance, image and status, attraction of more customers and users. There may be other indirect benefits such as the conservation of non-replaceable resources, waste reduction and sustainability; these may be further exploited by clients who market their facilities and infrastructure as environmentally friendly and green.

Crucial to the reliability of whole life cycle costing is the availability of sufficient reliable data to enable accurate appraisal of the performance and running costs of infrastructure. Performance can be measured against benchmarks or targets to assess usage, traffic flows, movements, wear, energy consumption, output and efficiency. This implies that systems should be set in place to monitor and report on performance. It will be part of the facility manager's responsibility to ensure that data generated is captured and processed to assist the establishment of performance targets intended to provide for the control, management and operation of infrastructure facilities (Armstrong, 2002).

Running costs should be categorised and where appropriate cost centres should be applied to provide more detail about precisely where individual costs are occurring. Figure 10.5 shows typical cost categories and related considerations.

Category of Running Cost	Consideration
Planned Maintenance and Renewals	Elements of the infrastructure should be designated a period for replacement or renewal. e.g. road surface will be replaced every 7 years. Planned maintenance will take place every 12 months subject to inspection.
Emergency repairs	Adhoc, but should be kept to a minimum
Cleaning	Regular schedules
Landscaping	Regular schedules
Fuel Charges	Oil and gas usage
Electricity	Lighting and power
Water and sewerage	Charges, maintenance and cleaning
Surface water disposal	Maintenance and cleaning
Security	Staff, CCTV, Alarms, barriers, locks etc.
Porterage and general staff	Staff loadings
Insurance	Loss, fire, third party liability, employers liability etc.
Taxation	Corporation tax, local taxation, employment costs
Management and Administration	Staff, accommodation, equipment, communications, transport

Figure 10.5   Categories of running cost.

Preliminary estimates of running costs can be prepared at the concept stage, but it will be necessary for the scheme design to be sufficiently worked up in order to prepare more detailed estimates of running costs. In the event that actual cost data are not available, estimates will need to be prepared using the best possible information sourced from manufacturers, suppliers and experience. Energy consumption will be calculated and estimates of usage will determine the need for repair and replacement. Costs should be calculated based on present values. Where data can be extracted from records generated by past projects this will provide a valuable source to reduce the risk of uncertainty associated with cost predictions.

Estimates should be made of intangible costs associated with the design. Such costs may be applicable to the whole design or only to certain parts. These may concern intangibles such as prestige, image or reputation. Unfortunately, these costs are difficult to estimate, but if in doubt it is best to tend towards being conservative.

Once all costs are calculated, the benefits should be identified and quantified. Benefits will be tangible and intangible, the latter being more difficult to quantify. This will then provide a basis for undertaking a cost/benefit analysis.

The next stage in the appraisal process is to examine the design to establish the feasibility of incorporating alternatives and assessing the effect on capital and running costs. Alternatives range from holistic design changes to changes in specific elements and these will provide the basis for various types of analyses.

Value analysis can be used to examine the potential for maintaining or improving functional performance by eliminating unnecessary costs and finding more efficient materials and methods (Dell'Isola, 1997). This requires elements to be disaggregated into their component parts and then each part is subjected to rigorous justification and exploitation to effect improvement. Value is categorised as aesthetic, function and esteem from which benefits can be derived and set against costs to produce performance ratios to measure improvement and make comparisons between alternatives.

Further analysis can be undertaken using discounting techniques where the selection of the correct discount rate is crucially important. Sensitivity analysis is used to measure the effect of incremental changes to one or more project elements. In this manner those elements with the most potential for gain can be identified and exploited. Probability theory and risk assessment can be used to gauge uncertainty attached to decision making. Financial techniques are covered in Chapter 7.

Life cycle appraisal (LCA) for PPP projects

A key requirement of PPP infrastructure projects is the knowledge and ability to visualise and predict the performance of proposed projects, in whole and in part. The selection of the final design must be in accordance with the appropriate choice of materials and components and their expected life and performance. Current knowledge of the impact of construction activities on the environment is perceived as becoming increasingly important. To this end, LCA has become internationally recognised as an all-embracing methodology. Several international agreements have been formulated under the ISO 14 000 series, which are aimed at a methodology that harnesses compatibility. In line with European directives, member state governments have placed great emphasis on the role of LCA as the gateway to an all-round remedy for all environmental mishaps.

The relevance of LCA relates to the proposition to provide the means whereby design decisions can be informed by the ability to visualise and evaluate life cycle outcomes relative to cost and performance (Alshavi and Faraj, 1995). In this manner the true life cycle effects of design changes can be determined and compared before a final selection of the optimal solution is made. Visualisation should be in 3D and ideally should be extended to include the fourth dimension of time. In this manner the physical form of the building can be viewed in simulation mode to allow designers, constructors and users to examine in more detail the characteristics of life cycle function and performance. This will enable more informed design decisions to be taken in the light of simulated performance of materials, components and elements, together with their running costs.

This approach implies the use of a model to replicate and simulate 'in-use' conditions experienced during the life cycle of an infrastructure project.

Normally infrastructure projects comprise one or more structures and facilities that are constructed on one or more individual sites. Each site will possess different characteristics which are likely to directly affect function and performance. The location and the availability of service utilities, e.g. electricity, water, sewerage and waste disposal may have a substantial impact on design. Planning restrictions and the proximity of other essential transport and urban infrastructure will also be an influencing factor.

Exposure conditions will have a direct influence on structural forces caused by wind, and where applicable, in combination with rain and snow. Climatic and atmospheric conditions throughout the annual seasonal cycle should be established, especially temperature changes between night and day, humidity variation and the nature of atmospheric pollution. The presence of pollutants will need to be assessed to aid the selection of materials that are best able to accommodate pollution present without premature discolouration, corrosion and deterioration in the form of decomposition, de-lamination, cracking and crazing. Visualisation must therefore be provided with sufficient environmental information to enable consideration of all possible circumstances, given the selection of a site in any geographic location.

The physical nature of an infrastructure project is derived from the design concept and the utilisation of technology in the form of materials, components and elements. Selected materials and components will be used in accordance with the design to create the major project elements, all of which will be connected to form the structure and where appropriate the internal environment. By utilising a construction method as defined by a selected work breakdown structure (WBS), human resources supported by equipment can be project-managed to enable construction. The WBS will be defined as a sequence of operations necessary to construct the project.

Components can be categorised as those constructed on site, *e.g. in situ reinforced concrete walls* and those processed off site, *e.g. steel frame components, cladding panels and door sets*. The former will require human intervention in the form of site engineering and quality control to ensure compliance with drawings and specifications. The latter will require fixing in place by human resources and equipment according to manufacturers' instructions and will invariably be dependent on dimensional accuracy of in situ site-constructed components and elements.

The total quality of the design and the construction process will have a direct impact on the efficacy of the completed project and its life cycle performance and associated running and maintenance costs. It is therefore vital that the design is fit for the purpose, as defined by the intended use of the infrastructure and its physical environment. Design failures can manifest themselves in many different ways; however broadly speaking, these may be categorised as aesthetic, spatial, structural, exposure, functional and incompatible component/material interaction. Construction process failures can be caused by poor workmanship, non-compliance with drawings, specifications and manufacturers requirements, together with non-conformance with construction sequences and damage caused by adverse environmental conditions. These failures can be largely avoided by tight project control and the implementation of adequate protection measures to counter environmental damage.

The quality of the completed project will have a direct bearing on its ability to meet and exceed the expected design life while maintaining full functional performance within anticipated running and maintenance costs. To assist architects, engineers and facility managers it is vital that they have the ability to visualise the life cycle effects and implications of alternative design options, given intended use and function, set within the context of the built facility's physical environment. The philosophy proposed is based on the development of a means to simulate the construction process and subsequently every aspect of the whole life cycle, using visualisation made available by the latest developments in information technology.

Proposed basis for an FM visualisation model

The ultimate goal of the model is to develop a holistic system for life cycle design and maintenance of infrastructure that utilises both immersive realistic visualisation and knowledge-based decision rules (Khosrowshahi et al., 2004). The idea is in itself innovative and the resulting product is based on a novel notion that it is possible to visualise the degradation of an entire facility throughout its life cycle as shown in Figure 10.6.

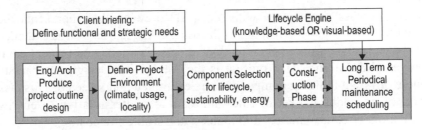

Figure 10.6   Model for life cycle design, repair and maintenance.
Source: Khosrowshahi et al., 2004.

The model must have the potential to generate a *just-in-time* maintenance programme for the project life cycle. It should offer an easy-to-use tool with a multi-view interface that is tailored to the needs of individual end-user. The decision component is based on the choice of either a knowledge-based or a visual system. The life cycle behaviour of project components is simulated in a virtual structure or facility and this knowledge is translated into a maintenance programme. In the visual environment, the observer should be able to make informed decisions about the timing of actions; but this process should also be automated through the use of a knowledge-based system.

The core technology of the visualisation model is its 'maintenance engine'. Its purpose is to assist the generation of *just-in-time* planned maintenance by mathematically simulating the behaviour of components over their life cycle and making data available through the visualisation mechanism.

The overall generalised model is given in Figure 10.7. There are three parts to the visualisation model, namely, data, simulation and presentation. The knowledge associated with the subject domain is encapsulated into data-objects that will be incorporated into a heterogeneous database containing component details, time-related behaviour and their visual attributes.

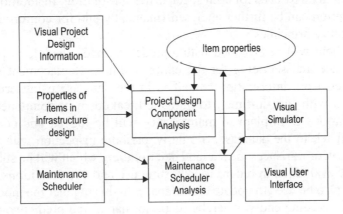

Figure 10.7   Generalised simulation model.
Source: Khosrowshahi et al., 2004.

273

The process commences with the CAD drawing that represents the initial state of the project. It contains the usual CAD data as well as extended attributes such as parameters, component specifications and project environmental definition. Also present is the Alternative Design mechanism where different building components are examined and selected. This task contains an additional set of rules and data attributes relating to energy evaluation, fiscal parameters and indices, and whole life costing.

The model proposed should use web capability to facilitate access by product suppliers and manufacturers who will provide data about materials and components. Designers, constructors and facility managers, who will be the beneficiaries, participate remotely by providing project information prior to participating in interactive analysis visualisation resulting in maintenance reporting. The holistic and integrative nature of the model is further enhanced through collaborative visualisation. The proposed model is intended to provide a host of hitherto unavailable facilities initially aimed at providing designers with the ability to visualise the consequences of design decisions that extend throughout the entire life cycle. Subsequently constructors and facility managers will be provided with post-construction maintenance and repair planning and monitoring.

Statistical analysis of data will be undertaken in order to develop mathematical representation of the lifetime status and expectancy of all items of the design. Once the behaviour of a component is statistically determined and its visual attributes are defined, then its physical appearance can be visually demonstrated through the proposed visual user interface system.

The proposed model is a sophisticated entity that provides a host of hitherto unavailable facilities initially aimed at providing designers with the ability to visualise the consequences of design decisions that extend throughout the entire life cycle. In this way, LCC can be determined to cover repairs and maintenance for each specific design solution. In addition, each design option can be further analysed financially and for conformance to sustainability principles.

The client will be provided with a predetermined schedule of planned maintenance and servicing for the entire building. Replacement will be determined from information contained in the database using principles associated with 'just-in-time', to mitigate breakdown or component failure. Variations from planned maintenance will be logged and converted into feedback to the database. Similarly, periodic inspection will provide information on variances in wear and deterioration, which will be subjected to further evaluation in regard to exposure, use and site conditions. These data will also be used to update the database. In this manner the model will become a learning entity, whereby its performance and predictive powers will be subject to continuous improvement.

Work will be required for the development of a knowledge-based broker based on a web-based data capture system that will enable manufacturers and suppliers to input their data into the model for conversion into knowledge. Other areas requiring significant generic research work will include the use of case-based and multi-agent reasoning techniques in order to generate the knowledge model that will underpin further development. Potential development and use of the model will not be limited to producing new knowledge. It will also initiate a change process by facilitating a smooth integration into the modern age whereby the design process will be more integrated, transparent, and reflective of strategic goals rather than the current practice of placing company strategy at the mercy of project objectives.

This section is intended to provide a glimpse at the future and what could be developed to give clients and infrastructure facility managers a higher degree of control over projects.

Operational facilities management

The operational organisation of facilities management will depend on the strategy selected. At one extreme will be the operation of all activities through an in-house division employing wholly direct labour and at the other will be the decision to outsource all operational and maintenance activities to a specialist facility management contractor. A mid-course action would be to appoint an in-house facility manager with the necessary managerial and administrative support who will oversee the appointment of a FM contractor to manage some or all service providers (Figure 10.8).

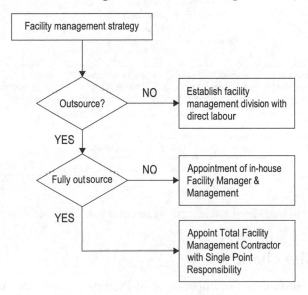

Figure 10.8   Strategic facility management options.

Direct labour organisations have mainly fallen out of favour for reasons stated earlier in this chapter and the degree of risk involved. There is also the need to divert energy and resources to facility management, which is not part of core business. Where it is desirable to retain an in-house FM capability a managing agent can be appointed to oversee and manage directly employed staff and external service providers. However, it is important that performance standards are properly specified and such contracts should be subject to periodic review. It may be more appropriate to outsource the entire FM function using an agreement with specified levels of service. This option has the advantage that total costs are known in advance and risk is devolved to the total facility management contractor. Alternatively, an in-house facility manager may be appointed with the necessary support to manage contractors who comprise the operational and management supply chain.

In the case of large-scale infrastructure distributed over a considerable geographic area involving several sites and linear service routes, e.g. rail track, expressways, power lines, etc., it may be necessary to devolve operations into regions, which in turn may be further subdivided into specific locations. This raises many organisational issues concerning the appointment of key managerial staff, administrative support and policy regarding outsourcing. Figure 10.9 illustrates a regionalised facility management organisation where the amount of decentralisation of management and decision-making needs to be carefully balanced against the advantages of systemised, centralised administration and support. The option exists to appoint facility managers at all levels as direct employees with permanent or fixed-term contracts; the alternative is to engage them as consultants.

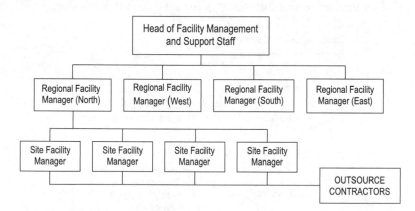

Figure 10.9   Regionalised facility management organisation structure.

Role of the client

The client's attitude to the importance of facility management in supporting the effectiveness of the core business is essential to proper implementation. In the past some clients have treated the management of facilities as

276

little more than essential maintenance and repairs. Therefore for facility management to be properly applied the client must

- be convinced that a total approach to managing facilities in support of the core business will enable greater efficiency and a better environment for users and employees; and
- be prepared to make the necessary investment to the proper implementation of FM.

The client must understand the needs to be imposed on facility management and should be able to specify service requirements and targets. In addition, the client must

- oversee the management of outsourcing with due regard to risk;
- agree on performance standards applied to in-house activities and outsourced services;
- approve and oversee monitoring and control;
- evaluate user/customer satisfaction levels;
- agree and approve changes to service requirements;
- oversee the ongoing development of new strategies according to the changing needs of the market;
- check that all statutory and legal requirements are being complied with, including health and safety; and
- ensure that sufficient investment is being made in innovation and training.

The client is the pivotal figure who sets standards and demonstrates leadership in the achievement of strategic goals (CIC, 2002b). The client's responsibilities will be delegated according to the organisation structure selected and will vest authority in individuals or teams to undertake the necessary tasks to achieve established targets. Key to this delegation will be an adequate understanding of accountability at all levels, which ultimately rests with the client. The development of good relationships with outsource contractors and members of the supply chain will emanate from the client and where possible partnerships should be built that are based on trust and understanding.

Packaging and outsourcing contracts

An objective appraisal must be made of those functions of facility management to be outsourced and those that will be kept in-house. In the event that the total FM option is not selected, there will be a need to retain direct control, thereby providing the client with the reassurance of onside advice and direction. There may be other key elements of the management process that may be undertaken in-house such as information, control and accounting systems intended to provide data to establish conformance with benchmarks and levels of service. Inspection of quality, health and safety will also be a prime candidate for direct control.

There may exist some crucial processes where functions are so vital to the core business that clients may decide that these should be provided in-house to achieve strict control over process, procedure, health and safety. This might also include fire fighting and control, internal environment control and monitoring.

Usually infrastructure projects require a wide range of facility management services that can be readily identified as essential to efficient operation. These services will need to be closely coordinated and controlled to avoid disruption, failure in performance and inefficiency. Contracts should be packaged according to trades and specialisms that equate integrated functions and processes that are identifiable as providing a discrete service. Furthermore, the packages must align with the range of expertise available from contractors.

Typical infrastructure outsourced packages might include

- General repair and maintenance
- Security
- Cleaning: interior and exterior
- Space
- Landscaping
- Signalling and traffic control
- Track repair and maintenance
- Ventilation and air-conditioning systems
- Servicing and repair of M&E equipment
- Turbines, boilers, generators
- Lifts, escalators and horizontal movement systems
- Roads, pathways, terraces
- Lighting: internal and external
- Painting and decoration
- Drainage, waste disposal and sewage treatment
- Water treatment
- Waste disposal
- Space utilisation and maintenance
- Communication services
- Transportation
- Catering Services
- Office Services

In order to make outsourcing packages more attractive to service providers, it may be desirable to combine or group two or more of the above areas, in part, or as a whole.

Service level and contractual arrangements

Where work is to be outsourced it will be necessary to produce a service specification that quantifies the acceptable standard of service required by users or customers and this will usually form part of the contract with the service provider. The service specification will lay down standards covering the following:

- organisation policy and procedures;
- requirements under the contract;
- statutory obligations;
- health, safety and welfare requirements; and
- processes and procedures to acquire technical standards set out in manufacturer's instructions, codes of practice and standards, e.g. Euro codes, DIN, BS, ISO.

The Service Level Agreement expands the service specification by setting down a framework consisting of performance standards and procedures for achieving them. These should be sufficient to develop detailed procedures as the contract progresses. More detailed consideration will be given to industry standards and accredited practices and procedures provided by manufacturers and suppliers. There will also be specific requirements necessary to support users and customers, e.g. response times, maximum periods for emergency repairs. A typical service level agreement will normally contain the following:

- Scope and range of services to be supplied
- Performance and quality targets
- Prices
- Time schedules and targets
- Resource requirements
- Procedures for communication and interaction
- Provision for change

Tried and tested standard conditions comprising the facility management contract should be used wherever possible. These should be periodically reviewed to establish that all standard clauses remain relevant. Where clauses are not appropriate, they should be amended or deleted altogether. It is essential that the form of contract should be capable of providing a legally binding contract with no loopholes, thus enabling the effective implementation of service specifications and service level agreements. Provision should be made within the contract for changes to be made in order to accommodate effective working practices that might not have been foreseen at the outset. Where changes or variations incur additional cost or require deletions from the work, provision should be made to make the necessary financial adjustments in the contract. Typical forms of contract are produced by the Joint Contracts Tribunal (JCT) and the Chartered Institute of Building (CIOB).

A further option is to use term contracts where contractors tender on the basis of a schedule of rates and hourly charges. Work is authorised as required and measured by a term contract surveyor and priced according to a schedule of rates. Overheads and profit are allowed at the term contract rates to calculate sums owing to the contractor and they are normally paid on a monthly or periodic basis.

It may be desirable to incorporate performance incentives based on achieving and exceeding specified benchmarks laid down by performance indicators. In this manner continuing improvement can be encouraged and achieved.

The next step is to obtain tenders that comply with the service specification and the service level agreement. Those eligible to tender may be drawn from a selected list of preferred contractors, or it may be open to public tender in response to advertisement. An alternative would be for those wishing to tender to pre-qualify by submitting a case against laid down criteria to determine a short list.

Where strong relationships and positions of trust have been built up on previous projects the client may wish to enter a partnering agreement, whereby standards of service already achieved provide a basis for work to be undertaken against mutually acceptable performance criteria.

If competitive tendering is the selected procurement method, then it will be advisable to ensure that all tendering contractors fully understand all the requirements. It may be an advantage to hold briefing meetings where the documentation and requirements are fully explained and queries from tendering contractors are answered. It should be made clear that the lowest tender may not be accepted given that other factors may be taken into account, e.g. consideration of quality of service offered and other advantages to the client. This should be followed by an adequate time for the tender to be properly prepared and submitted and a strict deadline should be specified after which tenders should not be considered.

Tenders submitted should be subjected to detailed scrutiny and checking. Where mistakes are discovered, the tendering contractor should be notified and asked to stand by the tender. From the tenders submitted it may be necessary to draw up a short list for further consideration. Where tendering contractors are drawn from an approved list this will be a relatively straightforward process and will mainly involve points for clarification. In the event of the short list being drawn from a process of open tendering, further investigation aimed at establishing the ability of the contractor to carry out the work in a satisfactory manner will be necessary. Price should only form one of the criteria for selection, since the prime objective will be to achieve best value. Quality of service measured against benchmarks will play an important role in the process. The ability of the contractor to fulfil the contract technically and financially must also be assessed.

Once the winning contractor has been determined, arrangements should be made for a pre-contract meeting to agree final details regarding the programme schedule and integration with other service providers. Clarification of procedures and practices will be finalised, together with arrangements for monitoring and control by the facility manager. Administrative arrangements will be made for regular progress meetings and reimbursement methods will be agreed for services rendered according to the conditions of contract.

Facilities condition management system

At the heart of FM is the system for recording the condition of infrastructure on an elemental basis according to data collected by means of periodic inspection. These data will then be compared with the service specifications and the service level agreements of the outsource contractors. Each contractor and team will be expected to undertake maintenance and replacements according to specified standards of service and to undertake emergency work as necessary within agreed time parameters. Where shortfalls in service are established, mandatory registers should be checked and those responsible should be required to immediately take whatever action necessary to rectify the situation. In cases where the necessary inspections and work have been carried out, as agreed, attention should be concentrated on the performance of materials, components and fabrication. Where necessary, manufacturers and suppliers should be involved to investigate and rectify performance shortfalls. Inspections should be undertaken to establish that cleaning and sanitation is properly expedited with follow-up procedures as previously described.

Where it is a statutory requirement to undertake periodic safety and health inspections, checks should be made to establish that registers are kept up to date and that faults and defects are reported and acted upon.

It is part of the facility manager's role to monitor the utilisation of space, output performance of equipment, energy consumption, waste and inefficient working in support of core business. Moreover, internal environment affecting air quality, lighting, heating, solar gain and noise should be constantly monitored and performance below target should be acted upon.

Ideally, the recording and monitoring system should use comprehensive software that is able to collate and analyse data to produce meaningful control reports that will form the basis for corrective action. The system should also provide an audit trail of elemental repairs that can be used in future to assess the performance in the use of the design. This will require integration with the financial management system, which ideally should form a module of the same control package.

Financial management and cost control

Operating costs may be defined as any costs associated with keeping an infrastructure facility in fully operating condition according to specified performance indicators or benchmarks.

281

At the design stage it is essential that all LCC are properly considered and evaluated according to the following headings:

- energy consumption broken down into lighting, heating, air conditioning and ventilation, plant and equipment, transportation and ancillary needs;
- planned replacement costs for infrastructure components and elements;
- planned servicing and maintenance;
- cleaning;
- water treatment;
- waste and refuse disposal;
- security;
- ancillary staff;
- ancillary disposables;
- health and safety (including first aid);
- equipment and fittings;
- emergency repairs and contingencies.

Typically budgets should be established for each of the above headings based on all available information and data, including past records, manufacturers' performance data, industry norms and benchmarks.

Service provider work packages will be periodically monitored according to budgets and where necessary work packages will be subdivided into smaller allocations of work for closer monitoring and control. Provision should be made to account for and keep track of authorised changes to the scope of work packages or the conditions attached to service level agreements.

A system of regular reporting should be in place to detect variations from the budget. In the event of negative variances, reasons should be sought for poorer than expected performance and appropriate corrective actions should be taken to halt or reverse adverse trends. Data derived from the cost control system should be processed to generate a database of information to assist with improvement in future performance predictions and forecasts.

To encourage continuous improvement, the desirability of including incentive payments should be investigated and evaluated. Where benefits are achieved, there should be a mutually agreed system based on key performance indicators (KPI) from which the benefits gained should be divided between the client and the service provider according to prior agreement.

Application of facilities management to PPP

Referring to Chapter 6, Section B2 of the proposed procurement framework provides for a fully integrated delivery system from the inception of infrastructure provision to final disposal stage of the life cycle. This grouping of procurement methods is ideally suited to PPP, where the client cedes the right to operate infrastructure facilities to a concessionaire who generates income from either user fees, tolls or rents. The basis of this agreement is the expectation that at the end of the concessionary period ownership and/or operational responsibility will be transferred back to the public client, or alternatively a further concessionary period will be nego-tiated. There are variants to PPP, the British version is known as the private finance initiative (PFI) and the World Bank system is known as private participation in infrastructure (PPI). Typically, concessionary periods are between 15 and 30 years and it is essential that risk is properly assessed and income streams are adequately predicted to help ensure that capital expen-diture is covered and that there is an adequate return on investment after all costs have been covered. Chapter 7 deals with the methods of assessing risk and the prediction of financial outcomes.

The concessionaire's responsibility for operating infrastructure, post-construction phase, requires greater consideration of in-use performance of the preferred design. This implies that a value analysis approach should be adopted throughout the design and construction phases to ensure opti-mised facility performance according to criteria laid down in the client brief. To undertake this task adequately, it will be necessary to be in possession of all the required information and data regarding operational performance and costs. Expertise will also be required to evaluate the util-isation of land, infrastructure facilities and space within buildings (McGregor and Then, 1999). The involvement of an experienced and expert facility manager at all stages of design will contribute to a more strenuous appraisal of performance and costs associated with operation. In this sense PPP has introduced new approaches requiring a greater degree of analysis at the design stage to improve in-use performance and reduce risk. During the construction phase competent project management control will be essen-tial to ensure the implementation of construction methods and sequences that do not compromise quality. Construction design management should be implemented to ensure safe and healthy designs, and it is recommended that a safety supervisor should be appointed to oversee design and con-struction (Akintoye et al., 2003).

Knowledge of life cycle performance of all infrastructure components and elements will require evaluation to form the basis for maintenance, repair and replacement budgets and sufficient contingency sums should be allowed to cover unexpected breakdowns and emergencies.

The amount of data processing and reportage can be particularly time-consuming and expensive. The use of suitable proprietary software will assist the effort and cost associated with this type of data processing and it is likely that the quality of control reports will be improved to assist more effective decision making. The potential to develop more sophisticated functions to improve efficiency and reduce in-use costs, especially at the design stage, is at a relatively early stage of development. The benefits to be gained from developments in this area have the potential to continuously improve future infrastructure performance while advancing in-use performance and quality with less client disruption and improved profitability with less risk.

Case studies

Case study 10.1: Equion FM Ltd.

Equion FM is an independent subsidiary of Equion plc, which in turn is owned by John Laing, UK. Equion concentrates solely on investing in PFI and PPP project companies that specialise in the areas of defence, health, education and emergency services. Equion FM works closely with it's parent company to manage property and facilities. The service offered to private and public clients is fully integrated and includes asset management with whole life cycle operations if required, a full design service, procurement, contractor performance, service level agreements and benchmarking. Other services include advice upon structural analysis, implementation of management structures, organisations, information systems and business processes.

Equion FM claims to have developed a distinctive approach to analysing FM system requirements in terms of inputs, processes, outputs and performance. Each project is subjected to cost benefit study and risk analysis intended to maximise benefits for clients and to minimise risk.

Current projects range from £4 to £500 million and normally range in duration from 15 to 35+ years. Advice is given to design and commissioning teams based on data derived from actual building performance benchmarks and experience gained in achieving service level agreements for new and existing built facilities. The overall company knowledge base provides effective management of assets and systems, and workspace can be designed and configured to facilitate new working styles and processes.

Equion currently holds ISO 9001 quality assurance certification for all operations and directly managed activity. It also has achieved accreditation according to ISO 14001 Environmental Management.

Author's commentary

Equion FM is an excellent example of a specialist facility management subsidiary that provides specialist FM services to its clients in close co-operation with its parent company Equion plc. This case study demonstrates the need to take full advantage of FM experience

*at the design and commissioning stages to improve built facility per-
formance and to provide an internal environment for efficient and
effective client functions. It also illustrates the need for financial,
legal, architectural and facilities management specialists to work
closely with clients to design, build, finance and operate new, refurbished or regenerated built facilities.*

Case study 10.2: Taylor Woodrow Facilities Management

Taylor Woodrow has developed its property management expertise
over the past 80 years and Taylor Woodrow FM now offers business
partnerships to provide bespoke programmes of strategically
planned care. PFI/PPP is incorporated where required to provide an
integrated approach to involving finance, construction and FM. FM
services offered include life cycle analysis, cost modelling, innovative
maintenance programmes, human resource management, proac-
tive and reactive maintenance and the provision of soft services. FM
involvement commences at the design stage and follows throughout
construction, commissioning and life cycle operations. The systems
developed by Taylor Woodrow FM ensure that built facilities perform
to their maximum potential facilitated by a framework of economy
and efficiency. In this manner all concerned are able to extract maximum value when the agreement reaches maturity and to the facility when it is handed back to the client.

Taylor Woodrow FM acknowledges the importance of the contribution of the efficient operation of property and infrastructure and
the impact this has on the profitability of the client's corporate business. This is particularly the case in the telecommunications sector
where Taylor Woodrow FM has been working with all major operators in the design, construction and ongoing maintenance of mobile
phone network infrastructure. In this area strategic partnerships
have been developed with companies that have complimentary
skills and knowledge, thereby providing greater client choice by
means of providing integrated solutions. Such solutions must be
economic, feasible and sufficiently robust to meet the requirements
of a technologically advanced and dynamic industry.

Taylor Woodrow FM has invested in the development of a multi-
million GBP system that accommodates FM typified by the need to
manage a high volume of low-cost transactions necessary for the
delivery of an integrated process-driven solution for telecommunication activities. Taylor Woodrow FM claims to have more than halved
operational costs for its clients, while at the same time adding value
through enhanced network reliability. It can also call on the full

resources and expertise of the entire Taylor Woodrow Group, if required, to further enhance the service provided to the telecommunications industry.

Author's commentary

This case study demonstrates how FM can identify and strive towards meeting the needs of a niche market. In this manner vital knowledge and skills are acquired through experience that helps to build competitiveness by means of a range of integrated and flexible services to meet specific customer requirements. Taylor Woodrow FM also provides services to other sectors, including healthcare and private corporate customers. It works in close association with other companies in the group and draws on PFI/PPP expertise and know how to provide holistic DBFO style cradle to grave solutions for its clients.

Summary

The efficient and sustainable operation and management of infrastructure to maintain function and performance throughout a specified life cycle is a key requirement against which all projects can be judged. The consequences of decisions taken by the design team are now recognised as having far-reaching repercussions that become more costly to put right as time progresses. This chapter highlights the importance of fully understanding the implications of selecting combinations of materials, components and fabrications that are brought together by the designer to form the prime elements required by the infrastructure to be provided. The involvement of an expert and experienced facility manager throughout the design is an important action that aims to provide the design team with a more informed knowledge of performance in use, upon which long-term decisions can be taken. This should be supplemented by a policy of collecting, analysing, storing and retrieving relevant information about the functional performance and other aspects of materials, components and fabrications utilised by infrastructure projects.

The delivery of infrastructure is portrayed as an integrated process where build quality is realised through the integrity of the design and the skills of the construction team. This requires designers and constructors to work in partnership in the interest of the client and the eventual users and customers of the infrastructure being provided. The facility manager should be fully involved in this process, such that a key contribution can be made to commissioning and handing over projects in a professional and seamless manner. In this way the facility manager should be fully conversant with all aspects and will have prior knowledge about the nature of the infrastructure. This will be invaluable in working up systems and procedures to their operational performance benchmarks.

The need to tightly manage the operation of infrastructure using in-house and outsourced services according to circumstances and need has been stressed. The management and co-ordination skills required of the facility manager have been identified as core attributes, together with the ability to anticipate problems and change. Success will be judged by the achievement of performance benchmarks and the safe operation of infrastructure within specified budgets.

It is clear that facility management has grown in stature and importance over recent years and indications are that it will continue to do so as the need to sustain and conserve non-replaceable resources increases. There is still much to learn from current and future innovation, together with creative solutions that will provide adequate sustainable infrastructure to meet the needs of clients, users and society. Most existing infrastructure provision is not sustainable and will eventually cease to function. The future challenge is to find new ways of designing, constructing and managing infrastructure.

References

Akintoye, A., Beck, M. and Hardcastle, C. (2003), '*International facilities and property*', Information Ltd, Blackwell Publishing, Oxford, UK.

Alexander, K. (1996), '*Facilities management: theory and practice*', E & FN Spon, London (ISBN 0-419-20580-2).

Alshavi, M. and Faraj, I. (1995), '*Integrating CAD and virtual reality in construction*', Proceedings of the Virtual Reality and Rapid Prototyping for Engineering Conference, Salford University.

Armstrong, J. (2002), '*Facilities management manuals – a best practice guide*', CIRIA, London.

Atkin, B. and Brooks, A. (2000), '*Total facilities management*', Blackwell Science (ISBN 0-632-05471-9).

Barratt, P. and Baldry, D. (2003), '*Facilities management: towards best practice*', 2nd Ed., Blackwell Publishing (ISBN 0-632-06445-5).

Construction Industry Council (CIC) (2002a), '*Design Quality Indicators*', online at www.dqi.org.uk.

Construction Industry Council (CIC) (2002b), '*How buildings add value for clients*', Thomas Telford, London (ISBN 0-7277-3128-9).

Dell'Isola, A.J. (1997), '*Value engineering: practical applications for design, construction, maintenance and operations*', R.S. Means Company, Kingston, MA.

Khosrowshahi, F.K., Howes, R. and Aouad, G. (2004), A building maintenance tool for PFI projects, 1st international CDVE Conference, University of Mallorca, pp. 213–220.

Langston, C. and Kristensen, R.L. (2002), '*Strategic management of built facilities*', Architectural Press, Oxford, UK.

McGregor, W. and Then, T.S.S. (1999), '*Facilities management and the business of space*', Arnold Publishers, London.

Nutt, B. and McLennan, P. (2000), '*Facility management – risks and opportunities*', Blackwell Science, Oxford, UK.

Park, A. (1998), '*Facilities management: an explanation*', 2nd Ed., Macmillan, London.

Spedding, A. (1994), '*CIOB handbook of facilities management*', Longman Scientific and Technical (ISBN 0-582-25742-5), Harlow, UK.

Chapter 11

Conclusion and future horizons

Introduction

The provision of sustainable physical infrastructure is a key factor in facilitating a higher level of socio-economic development. Fundamental to the delivery of physical infrastructure is the capacity to develop robust policies that can be effectively implemented in an integrated manner, utilising priorities set by need, benefit and affordability.

It is essential that policy and strategic frameworks are developed to take into account a long term view of the future and consideration of wider causes and effects well beyond boundaries normally associated with providing physical infrastructure. Such factors may include demographic trends, demand scenarios, the effects of global warming, economic trends and cycles.

Physical infrastructure supports growth and sustainable development by providing services to improve productivity and the enhancement of social exchange and accessibility (Swedish Ministry of Industry, 2002). By improving efficiency and reducing waste infrastructure contributes to national welfare and well-being. In this context, national welfare can be defined in economic, environmental, social and cultural terms. It is therefore difficult to quantify the full extent of services provided without taking a systemic integrated approach with the ability to relate the broader social and economic policies, while recognising cultural demands and influences.

A forward view is taken concerning the adoption of a policy framework and a systemic approach is proposed for its implementation. The sustainable use of resources, including renewable resources is examined and views are expressed on how future physical infrastructure provision should be developed and implemented.

The concept of service delivery is further developed taking into account the impact of public–private partnerships (PPP). Innovation is cited as being a key factor in achieving improved performance and the provision of services that are more efficient and provide better value. Future challenges are explored such as the use of viable systems models as a total approach to infrastructure provision and the development of new forms of energy to support sustainable infrastructure development.

An appraisal is made of innovative methods of providing infrastructure project finance by utilising local and national funding, debt finance and credit assistance. The chapter concludes with an appraisal of holistic risk management for infrastructure projects.

Physical infrastructure policy framework

Crucial to the development of a physical infrastructure policy framework is the influence of the environment, which must be understood in all its aspects, thus enabling the development of meaningful provision of infrastructure associated with the built environment. The prime elements concerning policy are the actors (institutions), resources, knowledge and information systems. These must be balanced against goals and objectives to facilitate implementation.

The scope of the infrastructure policy framework should incorporate both growth and sustainability over the longer term, taking into account externalities that have the potential to impact on the achievement of goals. It is therefore essential that the policy framework should be developed to include:

- Organisational and systems analysis to define the nature of infrastructure and to provide the basis for performance prediction.
- Cost–benefit analysis of improvements to correct weaknesses in existing provision.
- Monitoring of the influence of externalities in terms of opportunities and threats.

The development of policy should be in harmony with the prevailing economic climate; hence to perform adequately over the longer term there should be room for manoeuvre to accommodate varying conditions. Further, there should be a balanced approach to risk, taking into consideration those who should most appropriately carry all or part of the consequences.

The policy framework proposed should recognise the following principles, upon which policy can be predicated:

- Infrastructure policy is only one part of a set of socio-economic policies and should be in harmony with other macro objectives.
- Divestment of responsibility for physical infrastructure at the regional and local levels.
- Sustainable infrastructure should be geared to transforming resource inputs into improved and more efficient outputs that conserve resources and reduce waste.
- Infrastructure should be viewed as an operational systemic network incorporating all aspects of provision appropriately categorised into economic, social and trade infrastructure.
- There should be flexibility within the policy adopted to allow effective response to change.

- Trends affecting supply and demand for infrastructure should be continuously analysed and assessed, including the generation of gap analysis.
- Acceptance that there should be public or private ownership, or a sustainable public–private partnership according to need and circumstances.
- Performance should be continuously monitored and assessed against targets and benchmarks.
- Risks should be adequately assessed and a balanced view should be taken as to their distribution among stakeholders.

Figure 11.1 illustrates the main components influencing the development of goals and objectives, together with policy selection. Market supply and demand should be determined and related to the different categories of infrastructure. Relativities should be determined in terms of sustainability, resource utilisation and cost. These should then be related to policy issues and possible strategic actions and instruments should be established.

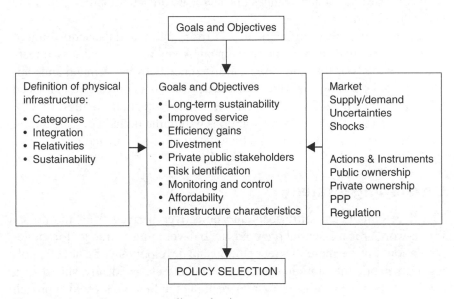

Figure 11.1   Influences on policy selection.

The major influences on supply and demand for physical infrastructure are predictable trends in demography and unexpected events. Demand is largely determined by the recipients of the service provided, who in turn are driven by economic and growth factors. Figure 11.2 provides a set of influencing factors that are categorised as major, secondary and sector specific. Most factors can be extrapolated to produce trends in demand. Unpredictable events are much more difficult to ensure suitable provision to mitigate causes and effects. Such events may occur at random or in clusters and in general they give little or no warning. In some cases, lessons can be learned from

Major	
• Demography	
• Economic growth	
• Industrial and commercial activity and dispersion	
• Occupation of premises	
Secondary	**Sector specific**
• Degree of employment	• Emphasis on fuel economy
• Density of building occupation	• Move towards an information society Income elasticity to satisfy the demand for transport
• Development of a service orientated economy rather that an industrial economy	• Tourism
	• Growth in international trade

Figure 11.2   Examples of factors affecting the demand for physical infrastructure.

studying history and past records, especially in the case of political developments and technological change. However, natural disasters, terrorism, conflicts, disease and sudden changes in fashion are difficult to predict with any degree of accuracy.

Key to the supply of infrastructure is affordability and the acquisition of the necessary funds to respond to demand. A key requirement is to undertake a gap analysis that addresses the difference between demand and supply and looks at ways in which infrastructure provision can be enhanced US Water forum (2003), US Environmental Protection Agency (2002). This brings into play innovative solutions within the domain of PPP, timing and size of investment, who pays, the quality of provision and regulation.

Developing strategy

The development of valid goals and objectives derived from the policy frame-work are an essential prerequisite to developing a strategy for implementation. Furthermore, the recognition and development of a holistic policy framework that anticipates causes and effects is ideally suited to a systemic planning approach. System mapping facilitates a layered approach where at the highest level macro elements can be defined, positioned and linked into an integrated framework showing key causalities and effects. Checkland (1981) advocated the use of 'Rich Pictures' and later Flood and Jackson (1988, 1991) proposed the use of mapping. The strength of this approach is that successive layers can be arranged in a hierarchy to provide a detailed view of subsystems that can be related together in network form that simulates the real-world activity (Howes and Tah, 2003). Different solutions can be tested using an iterative process to optimise and manage solutions. This is described in more detail in Chapter 4 where hard variables are defined as design, technology, infrastructure type, investment levels, location and implementation period. Soft variables are defined typically as institutional choice and influence that can alter the way in which infrastructure projects are procured, funded, designed and constructed. The treatment of soft variables accepts that these are simply best guesses but this opens up the possibility of what-if scenarios and probable outcomes associated with a range of possible solutions.

A total approach to the provision of physical infrastructure

A total approach for the provision of infrastructure involves institutions/ organisations concerned with each category of infrastructure provision and how they function. Assuming that an infrastructure category is treated as a subsystem, then a complete set of subsystems will represent holistic infrastructure provision. The total system of infrastructure comprises a complex amalgam of resources and technologies that are influenced by socio, economic and political factors set within an environmental context. The additional dimension of time introduces dynamic change that requires management and control.

By taking a view involving the adoption of cybernetic principles associated with the notion of 'viable systems diagnosis' it is possible to develop a model comprising of subsystems that defines an organisation as a total system and provides it with identity (Beer, 1985). By adopting the principle of recursion the whole system is replicated in its parts so that the same viable system principles apply throughout. Furthermore, operational subsystems are integrated and controlled through a higher level of management subsystems, taking into account influences from the environment. Emphasis is also placed on learning from experience and dynamic change with a view to the creation of new knowledge and skills to effect continuous improvement. This places importance on the need for intelligence gathering, auditing and feedback.

Building on the work of (Flood and Jackson, 1991) a 'viable systems' model (VSM) is proposed for the purpose of addressing the need for a structural system to support and promote sustainable physical infrastructure. Using this approach gaps and weaknesses can be exposed and the robust nature of the total system can be judged when subjected to trend scenarios and unexpected events.

Figure 11.3 illustrates a VSM that identifies institutional/organisational infrastructure categories as subsystems located in an operational domain. The operational functions are illustrative of the prime categories of economic infrastructure, which in practice will require expansion to include social and trade infrastructure. Each subsystem will be autonomously managed and controlled according to the traditional system concept of input–transformation–output that is subjected to influences from total and local environments. The model provides four high-level functions acting as subsystems that form a metasystem responsible for oversight, regulation,

293

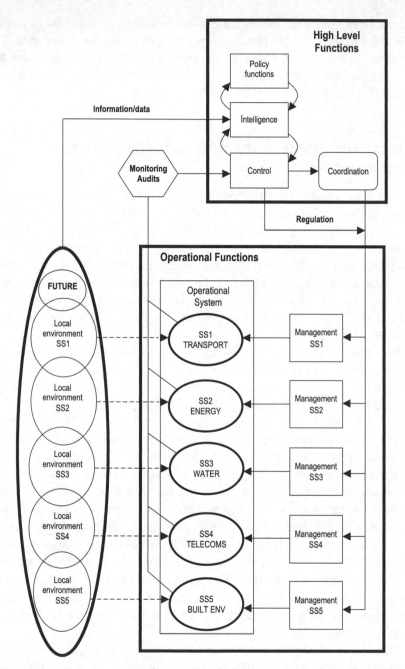

Figure 11.3   VSM for physical infrastructure.
Adapted from Flood and Jackson, 1991.

coordination and direction. The metasystem provides high-level thinking resulting from gathering intelligence from the total environment and audit reporting on the performance of the operation subsystems, both individually and collectively. Information collected is analysed and evaluated to

294

establish what actions should be taken and to initiate learning processes where the body of corporate knowledge is enhanced. The outcome of this process may result in policy changes at both the meta and operational levels to effect improvement as part of a continuing process of enhancement.

It is proposed that the VSM has direct relevance to national and regional governments who are seeking to provide a holistic approach to the provision of physical infrastructure that takes into account environmental, social, economic, political and social factors. In a report by the New Zealand Institute for Economic Research (NZIER, 2004) a template was proposed, as illustrated in Figure 11.4 that facilitates the reporting of monitoring using the operational function and the four high-level VSM functions set against infrastructure categories. When the template is completed it will highlight

Physical Infrastructure Institutional Breakdown	Operational				Coordination				Control/Audit				Intelligence/ data				Policy			
	So	En	Ec	Cu	So	En	Ec	Cu	So	En	Ec	Cu	So	En	Ec	Cu	So	En	Ec	Cu
Transport*																				
Energy*																				
Water*																				
Telecoms*																				
Built Environment																				

Figure 11.4   Template for institutional analysis.
Source: Adapted from NZIER.
(*) Further breakdown if required e.g. roads, rail, airports, seaports, etc.
So: Social
En: Environment
Ec: Economic
Cu: Cultural

weaknesses and gaps, as well as providing a basis for comparison between infrastructure subsystems.

The VSM provides a high-level overview, but for it to be effective it must be supported by systems that can be harmoniously coordinated to provide more detailed control systems within categories of infrastructure and at the project level.

In Chapter 4, an Infrastructure Organisation, Resource Management, Evaluation and Development Model is introduced (InfORMED). The InfORMED model uses a systematic approach to encapsulate the dynamic interaction between hard and soft variables to inform policy-making and strategic planning. It provides a policy component that determines the relative importance of policy variables and their impact on project investment policies. The policy component reflects the dynamic soft variables and analytical hierarchy process (AHP) is used to set priorities. The strategic component utilises an infrastructure resources model supported by hard project/programme and resource variables that interact horizontally with soft institutional variables. In this manner, the model enables infrastructure projects to be managed in a way consistent with policy priorities and resource constraints. The InfORMED model is more fully described in Chapter 5.

InfORMED provides a tool for high-level management that is complimentary to the VSM approach previously described. It is supported using Criterium DecisionPlus software and it provides another approach to evaluating policy and strategy in addition to considering detailed project specific data that forms the basis for resource management.

Sustainable infrastructure

By definition the Earth's non-replaceable resources are finite, therefore a dual approach must be taken to sustainability. On the one hand is conservation and on the other is the development of new technologies to take advantage of new materials and energy from replaceable sources. Both approaches need to be considered in the light of pollution and the consequences of global warming. There is also a need to raise general awareness of the adverse implications on future generations of today's consumption of their heritage through irresponsible and uninformed use and wastage of resources that cannot be replaced. This requires social and cultural change, which is the responsibility of the World's governments to bring about a greater sense of social responsibility for the well-being of the Earth's habitat and its population, both now and in the future. Changing social attitudes and cultures are normally slow processes and there must be a dedicated approach aimed at changing social attitudes and responsibilities. Over the past 50 years, anti-smoking campaigns typify how difficult this process can be and they are indicative of the level of effort required.

There is general recognition in developed and developing countries that research into the development of new technologies is now essential to provide new materials and sources of energy that are sustainable and non-polluting. This is an ongoing and expensive process. The government of the United States in its Hydrogen Posture Plan (2004) has declared that dependence on fossil fuels as the primary energy carrier is now seen as a threat to national security and plans are in the process of being implemented to attract over US$ 1 billion to support the research and development necessary to produce alternative energy sources.

Conservation policy

A multifaceted approach should be taken to conservation policy that identifies and implements a number of unified actions to reduce consumption and waste of resources. The recycling of non-replaceable materials linked to waste reduction is a key element, which needs to be viewed in the light of life cycle considerations and refurbishment rather than replacement. Innovation in design also has a significant role to play in reducing consumption and increasing performance and efficiency. A summary of recommended actions are as follows:

- Reduction in the amount of waste being consigned to landfill. This implies that better ways must be found for categorising and sorting waste in an efficient and financially viable manner in order to encourage manufacturers to make use of recycled materials and products to produce new materials. For example, old concrete should be crushed and graded for use as fill and reinforcing steel should be segregated and recycled as scrap metal. Efforts must also be made to increase the re-use and refurbishment of false works and at the last resort to increase the potential for recycling.
- The quality of design should be developed to achieve longer useful life cycle performance by reducing unnecessary wear and tear, minimising obsolescence and reducing the need for costly maintenance and repair. Potential should also be exploited to extend life cycle periods and where possible to facilitate refurbishment to extend the life of infrastructure. Improved-design should also play a part in reducing energy consumption and conserving water.
- Waste materials and by-products must be further exploited to produce new materials e.g. compressed fibre boards, composite materials, insulation and resilient sheeting.
- Steps must be taken to reduce the need for large amounts of embodied energy to be used during manufacturing processes. Where large consumption of energy is unavoidable e.g. as in the case of steel manufacturing and power generation then steps should be taken to use the heat produced for more than one purpose, rather than allow it to simply dissipate into the atmosphere. There are early examples

297

where power stations located in urban areas have been used to provide district heating and hot water for domestic purposes.

- Transport systems must be designed to seek maximum assistance from gravity, wind power and tidal flows. A simple example can be cited where metro tunnels have been bored to allow for down grades when leaving stations and upgrades on approaches. In this manner, the energy needed for acceleration and braking of trains is reduced. Other examples include modern sails to assist sea-bound transportation, tidal scheduling and the use of airships to transport heavy and bulky loads.
- Attention should be given to reducing the need for energy consumption by more efficient propulsion systems for vehicles, low-energy equipment, appliances and lighting. A major contribution is foreseen to be made from condensing boilers and passive ventilation systems designed to reduce the load on air conditioning systems.
- Smart control systems have the potential to reduce wasted energy.
- Reduction in the consumption of clean water by the prevention of leakage in the distribution system and at the point of use, together with the encouragement of local rainwater collection systems and the recycling of grey water.
- Regeneration schemes and plantation policies to provide continuous replacement for naturally grown materials, e.g. softwoods and hardwoods.
- Composting schemes aimed at ensuring the regeneration of top soil and the growth of plants and trees that act as sinks for carbon absorption from the atmosphere.
- Urban and rural land use strategies to reduce the need for unnecessary transportation.

It is a popularly held misconception that there is an infinite supply of natural resources that can support indefinite growth, but there is a growing awareness that this is not the case. This in turn has resulted in the belief that conservation places restraint on growth. Conservationists tend to refute this notion and instead promote the argument that growth and conservation are not at odds with each other. It is claimed that conservation can support growth by greater efficiency and benefit brought about by better design, processes and systems. Advantage is also claimed from improvements brought about by a greater public awareness of the need to reduce waste and to recycle. Moreover, this places dependence on the level of public awareness and greater social acceptance of the need to conserve resources and to draw attention to the proper management of fragile replaceable resources that could potentially become non-replaceable. Governments have a crucial role to play in creating public acceptance of the need to conserve non-replaceable resources and to support policies to recycle and to generate new technologies to exploit new energy sources and new materials that are replaceable.

Developments in the production of energy

Fossil fuel energy sources have two prime drawbacks, the first being that these will eventually be exhausted and secondly that they are responsible for the majority of the Earth's atmospheric pollution, which according to powerful evidence is causing global warming. The Kyoto Protocol has only recently been ratified and there is a growing realisation that financially feasible new clean energy sources need to be further developed. This section concentrates on existing and future sources of energy that with investment can conserve and hopefully eventually replace dependence on fossil fuels.

Hydrogen

The development of hydrogen as an energy carrier has the advantage of providing a long-term solution, since it is an abundant element and it will mitigate the effects of air pollution and greenhouse gas emission. The US Department of Energy has produced a Posture Plan in accordance with the National Hydrogen Energy Road Map released in November 2002 laying down provision to accelerate the development of hydrogen fuel cell (HFC) technology and its associated infrastructure. The programme sets down four phases and spans over a 40-year development period (as shown in Figure 11.5). The initial Research and Development phase will be followed by bringing hydrogen to the market and investing in the necessary infrastructure to commercialise this energy source. The plan is to invest US$ 1.2 billion over a 5-year period with government taking a strong R&D role for the first 15 years, after which industry will take on a strong commercialisation role.

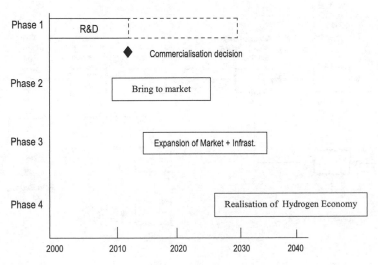

Figure 11.5   Hydrogen development timeframe.
Source: US Department of Energy.

The key objectives of this programme are as follows:

i. On-board vehicle hydrogen storage systems with a 9% capacity by weight to enable a 300 mile (480 km) driving range.
ii. Hydrogen produced from natural gas or liquefied fuels.
iii. Polymer electrolyte-membrane automotive fuel cells.
iv. Zero emission coal plants that produce hydrogen and power with carbon capture and retention.
v. Hydrogen production from wind-based electrolysis.
vi. Hydrogen fuel delivery technologies.

The elements of infrastructure required to produce hydrogen from bituminous coal are shown in Figure 11.6. This process uses advanced gasification technology, advanced membrane technology for hydrogen separation and CO_2 removal and carbon sequestration. An advanced two-stage clean up is used in combination with a ceramic membrane system operating at in the region of 600°C. The hydrogen produced is separated at low pressure and then compressed. The remaining synthesis gas is then combusted with oxygen in a gas turbine to provide power for the plant. Oxygen is used to produce a concentrated stream of CO_2 for sequestration. Heat recovery steam generators are used to produce steam, which is then sent to a steam turbine to produce additional power. It is anticipated that in future more advanced hydrogen and electricity co-production plants could provide significant additional reductions in the cost of producing hydrogen (Gray and Tomlinson, 2002).

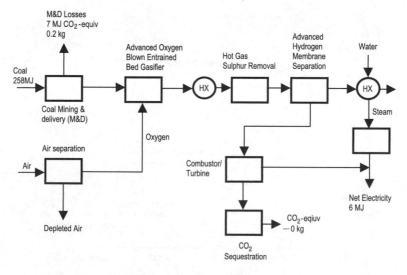

Figure 11.6   Production of hydrogen from bituminous coal.
Source: US Department of Energy.

There are three prime drivers that support the delivery of a hydrogen-based energy system. These are:

(i) *Energy security*. There is a growing World demand for fossil fuels, primarily in the form of oil, the largest reserves of which exist in the Middle East. It is anticipated that demand will rise by at least 50% by 2025. Hydrogen can be produced from a range of fuels, including renewables, nuclear and more readily available stocks of coal.

(ii) *Environmental cleanliness*. Atmospheric pollution accounts for poor air quality resulting in respiratory diseases, global warming and climatic disruption and disturbances.

(iii) *International competitiveness*. The major developed countries recognise the potential to free markets from over dependence on oil and the economic advantages that can be gained from the development and use of fuel cells.

HFCs directly convert chemical energy into electricity provided that they are supplied with fuel. HFCs have the potential to provide a 50% reduction in fuel consumption compared with gasoline and to increase the reliability of national grids by reducing system loads and bottlenecks. They also increase the cogeneration of energy in combined power and heat applications for buildings. Hydrogen energy will produce zero to near zero harmful emissions from vehicles to power plants.

Tidal energy

Countries with long coast lines and large tidal ranges have the most potential to benefit from tidal energy generation. In the United Kingdom it is estimated that if all the exploitable estuaries were utilised up to 15% of current electricity consumption could be produced from this source. The concept is based on the construction of a barrage built across all or part of a sufficiently fast-flowing tidal estuary, river or lagoon. The barrage incorporates turbines that are rotated by the force of the flood and ebb tides, thereby generating electricity. Power production is not continuous because the tidal flow decreases to zero at the point of high and low tide. However, it should be mentioned that the time of high and low tides varies around the coast of the UK and by strategic positioning of tide generators, the flow of electricity into the national grid could be evened out.

The Seaflow project in Western France consists of a barrage that also serves as a highway between St. Malo and Dinard. The barrage uses 24 No. 5.4 m diameter bulb turbines rated at 10 MW that annually generate 640 million kWh, equivalent to the needs of a city consisting of 300 000 people.

Wave power generation

Energy harvested from ocean waves potentially provides an unlimited source of renewable energy. Wave machines are designed to extract power from movement created by waves along shore lines or by deployment in deeper waters.

The Pelamis Sea Snake project developed by Ocean Power Delivery is a deep-water grid connected trial. When floating in the ocean the hinged joints between its cylindrical sections move with the waves, powering hydraulic motors to generate electricity. Each Sea Snake generates approximately the same power output as a wind turbine and the intention is that a wave farm of 1 sq km will supply enough electricity for 20 000 homes. This means of energy generation is still to be fully researched and exploited.

Solar power

Solar power provided by photovoltaic (PV) cells and heat pumps is ideally suited to areas that enjoy long hours of sunshine throughout the year. However, a major drawback has been the cost of manufacturing PV cells and this determines the relatively high cost of generating power in the region of 1 Euro/kWh. Currently there is some indication that costs and prices of active solar cells are falling, but there is still some way to go, even with generous government subsidies.

Ideal applications include power for remote telephone booths and roadside monitoring systems and in some cases they are used to supplement conventional heating and hot water systems on a small-scale basis.

Heat pumps are used in conjunction with strategically placed panels to raise the temperature of water by direct exposure to sunlight. The problem with these systems is that they become inefficient in cloudy conditions and they do not work at night. They are also susceptible to damage by frost, unless properly drained in low-temperature conditions.

Biomass and bioenergy

Biomass can be any type of organic matter located within the layer of living systems around the World known as the Biosphere. Bioenergy describes the various ways of generating energy from biomass. Examples of bioenergy fuels include waste, methane from landfill, sewerage effluent and household rubbish. There is also a recently developed technology termed as gasification, where solid material is converted into gas to drive engines and turbines to create electricity.

The arguments in favour of bioenergy are:

- Each unit of biomass directly replaces one that would otherwise be used to generate power from fossil fuels.
- Waste methane is captured instead of being released into the atmosphere where it adds significantly to the greenhouse effect.
- The provision of a potential source of income from rural areas.

Wind power

The generation of power from wind farms and single turbines has been covered in Chapter 9 and attention has been drawn to the considerable investment that is being made in parts of Europe where wind profiles are sufficiently advantageous.

Wind electrical power is generally proportional to the wind speed cubed, therefore if the wind speed doubles then the power generated rises by a factor of 8. The problem is that in very light wind conditions no power is generated and back up systems are required. Conversely, a 1 MW turbine will produce power for approximately 300 homes and will save in the region of 2000 tonnes of greenhouse emissions per annum. The cost of producing electricity by wind power is still relatively high when compared to fossil fuels and the incentive to invest in wind power lies with governments who are prepared to provide subsidies or tax breaks.

Summary

It can be deduced from the above appraisal of sources of energy that the only major development currently on the horizon with the potential provide for growing energy requirements is hydrogen technology. All other methods cited are relatively small scale by comparison and in many instances, as has been shown, are very dependent on favourable year round local conditions.

Taking a more positive attitude it would appear that collectively energy produced from water, wind, solar and biomass sources have the potential to create substantial amounts of electricity. This depends on the correct selection of generation that suits local conditions and the development of technologies capable of extracting more power output from renewable sources. Therefore, there is a case for investing in future R&D to unlock the potential for sustainable energy production.

The concept of service delivery

The introduction of PPP has helped to raise public expectation of amenity and services received from new and improved infrastructure. The impact of PPP is immense when one considers the scale of take up across the world and it is established as a viable means of providing, hitherto unaffordable infrastructure.

It is now appropriate to take account of what has been achieved and what should be done to seek continuous improvement in the future. The next step should be to improve the successful delivery of infrastructure projects and to accelerate the process of procurement, which at the moment takes too long (Wall, 2003). The imperative must always be to obtain the best possible service from infrastructure investment that has the capability to remove impediments to productivity and business enterprise.

The adoption of a service delivery approach in conjunction with PPP forces public authorities to define more precisely their requirements in terms of output based specifications and performance benchmarks in such a manner that will improve procurement performance and project delivery. This requires bringing together companies in consortia capable of bidding for the design, build, financing and the long-term operation of infrastructure over a concessionary period, typically between 15 and 30 years, as described in previous chapters.

The responsibilities of the consortium are such that design, in all its aspects, must be of the highest quality, not only to meet client and user output specifications, but also to ensure low operational and maintenance costs. The incidence of less failures and disruption to users helps to improve service delivery of the functions and processes supported by infrastructure. Crucial to the achievement of design quality will be input of expertise and 'know-how' from constructors and facility managers early in the design process. Input from specialists e.g. environmental engineers will also contribute significantly to improving functional performance and sustainability.

In addition to providing physical infrastructure, the level of services transferred from public to private sector control is very significant and often controversial. This is especially the case where existing public staff are transferred to new private employers. There are many issues surrounding staff transfers, including changing conditions of service, rates of pay and previously held restrictive practices that may be unacceptable to the private sector. Critics of the PPP-service-delivery approach argue that efficiency improvements are largely brought about by driving down conditions of employment.

However, it is clear that service delivery associated with infrastructure implies changes and reform in the delivery of public services. It is the duty of governments to act entirely in the best interest of the public and PPP projects rely on agreements and contracts alone for this purpose. Under normal circumstances this is straightforward in the procurement of physical infrastructure, but the delivery of key public services such as complex clinical care is likely to be highly problematic. Private companies are driven by shareholder pressure in the expectation of a decent return on investment and where this does not occur then arguments over conditions of contract could potentially be catastrophic to the beneficiaries of the services provided. Therefore in future the extension of PPP to deliver a wider range of key public services is a matter that requires great care and consideration.

An innovative development by the British government is the introduction of Public Investment Companies who are not owned by shareholders. Instead they are owned by various stakeholders, including government, the local community and other interested parties and it is possible that under certain conditions this might provide an alternative to PPP. As a case in point, the problems associated with the privatisation of the railways in the UK are cited. In this instance, the responsibility for the operation of the national rail network was divided between Railtrack, a private company who provided the permanent way, signalling and support systems and various private train operators who provided rolling stock and commercial systems such as ticketing. Due to the existing poor state of the railway infrastructure considerable investment was required and this eventually led to Railtrack being declared bankrupt by the government who then was obliged to take control in the public interest. The solution was to create a new private company called 'Network Rail' that would be financed by debt, rather than a mixture of debt and equity. Network Rail is governed by stakeholders namely train operators, government and public appointees. It is claimed that this overcomes the conflict of interest associated with shareholders. The downside is that considerable public investment has been necessary to upgrade the rail network that had become dangerous due to a lack of investment in essential maintenance and repairs, brought about by the policy to create a privatised rail network with split and overlapping responsibilities. Clearly this was a recipe for disaster. This case demonstrates the need for PPP schemes to be properly thought through and critically assessed from the outset to establish the benefits to be derived from specified achievable and sustainable outputs.

For the future PPP infrastructure projects will be governed by success in performance delivery over the service phase of the whole project life cycle. Therefore, design for serviceability from the outset will be a key factor to project feasibility and the benefits to be derived. This is a relatively new concept that has been gaining strength and significance as the level of PPP activity has increased and it is one that the construction industry must deliver successfully through partnership and integration of all parties concerned.

Innovation in project finance

There is an increasing international demand for physical infrastructure across developed, developing and least developed countries to support growth and public demand for improved services and facilities. It is now commonplace that traditional government funding sources are proving to be inadequate to meet increasingly diverse and in some cases complex needs. Therefore, the gap between the demand for infrastructure and what can be afforded from the public purse represents a constraint to economic growth and presents a challenge to find new funding techniques that complement and enhance existing funding mechanisms (FHWA, 2004).

There is no universal method for funding different types of infrastructure project, since some projects with high use volumes attract substantial revenues, while others may provide essential services for local communities but have little or no potential to generate significant user revenue. Hence, it is important to classify projects according to whether they are funded from private or public sources as shown in Figure 11.7.

The base of the pyramid represents the majority of infrastructure projects that rely upon grant-based funding because they do not generate sufficient user income, but instead provide essential services to support economic activity and the quality of life experienced by the populous. The middle section represents those projects that are able to attract both public and private

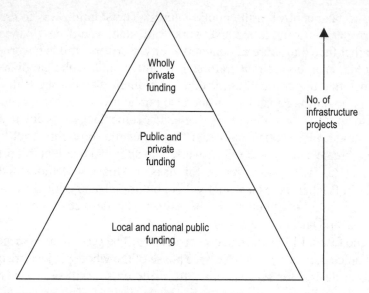

Figure 11.7   The infrastructure funding pyramid.

funding and may generate significant user revenues enough to provide partial funding, but will also require some form of public financial assistance to be viable. The apex reflects the relatively small number of projects that are capable of gaining private capital financing without the need for recourse to government support and assistance. Invariably in such cases the revenue generated will be sufficient to cover all capital and operating costs and will provide a return to investors.

The overall objective is to refurbish, upgrade or supply new infrastructure to meet demand and thereby remove constraints on growth as well as providing improved services. The task is therefore to provide the necessary means to leverage in funding in such a manner as to bring infrastructure projects forward even when insufficient funds are immediately available to complete the projects in hand. This approach requires the establishment of sufficiently robust criteria to ensure that projects are adequately considered and prioritised according to benefit and need. The remainder of this section considers the various alternative innovations to provide project finance.

Local and national public funding

The overriding aim is to provide local and national government with innovative financial management techniques to ease constraints on infrastructure development and to increase flexibility regarding the timing of obligations and reimbursements. Under these circumstances the maximum funding obligation of national government should be established in relation to that expected from local government sources. This will depend on the extent to which public funds are devolved to local spending control. Typically, a maximum ceiling provided by national government might be as much as 80% of the total project cost. The following key mechanisms are suggested:

- Projects should be allowed to commence even if there is insufficient public funding immediately available to cover all project costs. However, national government should commit to allocate funds in project stages. In this manner, the total budget available can be better balanced across a number of projects. This means that more projects can be started. Furthermore, where projects are able to develop a revenue stream, then this can be used to improve cash flow and thereby improve infrastructure development potential as a whole.

- Project financing can be tapered to provide a greater percentage of national funding available to be drawn down early in the project. Moreover, when user revenues are generated this assists cash flow, but the proviso must be that the local government funding element should be paid before the end of the project.

- Arrangement should be made to provide flexible local government funding. To make things easier for local government, it may be possible to make the local (match) funding requirement less arduous by not insisting on strict cash contributions. Instead, other sources could be taken into account, including donation of land, materials and services. This approach allows for accelerating the commencement of projects, releasing funds for further projects and promoting PPP by providing incentives to seek private donations.

- User charges take a variety of forms. In the cases of energy, water services and telecommunications, users will be billed according to their volume of use and a tariff of charges. There may also be a case for increasing charges to reduce consumption to mitigate the depletion of non-replaceable resources, pollution and the release of greenhouse gases. Charges for transport systems will be based on use e.g. fares, tolls, etc. or indirectly through taxation on fuel and the cost of licences. In the case of road transport there is the possibility of introducing toll credits where the revenue generated from toll charges can be used to support capital expenditure in the form of local match funding contributions. In this manner, more highway facilities can be provided and improved.

The principle behind the above suggestions is the facilitation of easier and less arduous financial requirements that are designed to enable local government in partnership with central government to bring forward new services and facilities, according to need and benefit, at the earliest possible time.

Debt finance

In the case of large infrastructure projects, local government may consider the possibility of borrowing by means of issuing municipal bonds that yield interest to investors. Such bonds may also yield interest that is tax free.

The cash generated by bonds enables infrastructure projects to be brought forward earlier than would otherwise be the case. Hence, the cost of borrowing can be offset against inflation, waste caused by congestion, poor services and deferred economic development. There may be other benefits such as safety improvement and reductions in pollution and greenhouse gas emissions.

Invariably, the repayment of bond financing requires revenues generated from services provided.

In the USA, Grant Anticipation Revenue Vehicles (GARVEEs) have been introduced. These allow states to pledge a share of Federal highway funding towards the cost of borrowing on a long-term bond issue.

Credit assistance

Credit assistance may be given to local government from central government in the form of loans or funds that are made available on a stand by basis in case revenues do not work out as expected. Credit enhancement (loan guarantees) reduce the risk to investors and enables the project sponsor to borrow at lower rates of interest. In this manner, project feasibility is easier to justify and schemes can be brought forward with consequential advantages, as previously cited, that can help to offset the cost of borrowing.

In the event that projects generate better than expected revenues it may be the case that a proportion of allocated national funds are surplus to requirements. In this situation flexibility should apply to enable, if necessary, these funds to be channelled into other projects provided that they fully comply with laid down funding criteria.

In suitable circumstances it may be possible to set up a regional or state infrastructure bank (R/SIB) whose role would be to make loans and guarantees for infrastructure projects. Figure 11.8 illustrates the basic structure of such a bank.

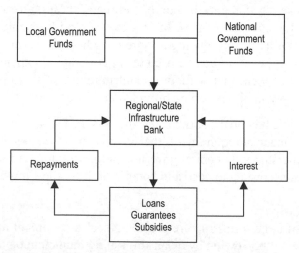

Figure 11.8   Basic structure regional/state infrastructure bank.

R/SIBs can be leveraged or un-leveraged. In the former case, bonds can be issued against the Bank's initial capitalisation, thus substantially increasing funds available for loans. Instead of loaning local and national government funds, these funds can be pledged as guarantees, together with the potential of anticipated loan repayments as applicable and used as security for a bond issue. The proceeds from the debt issuance can then be used to provide project sponsors with either loans or credit enhancement. An un-leveraged R/SIB simply lends available funds or provides credit enhancement for projects. Loan and interest payments are then recycled to fund future projects.

Bonds and loan guarantees are important vehicles to promote private investment in infrastructure, thereby facilitating a combination of public and private cash. This has the advantage of significantly increasing available funding for infrastructure projects, or it reduces the need for public funds.

Risk management of infrastructure projects

The quality and comprehensiveness of risk management plays an essential part in the successful completion of infrastructure projects. For risk management to be effective it must be viewed as an ongoing dynamic process from project inception to the termination of the whole project life cycle. As the project moves through its distinct phases it will be subjected to dynamic change from within and from its external environment. This will require correct and timely decision making at all levels of the project management structure aimed at achieving project targets and objectives.

The risk management structure should embody a framework that allows risks to be identified in good time to permit adequate analysis and evaluation (Lewis and Mody, 1998). The framework should also have a closely integrated risk monitoring system that is capable of identifying change and trends, together with the facility to assess likely impacts and causal effects. This should be integrated with management decision making to, either treat, mitigate or lay off risk (Conrow and Carman, 2000). A full understanding of the nature of the project and its environment is essential to identifying risks in a holistic manner, whereby the full implications of cause and effect can be assessed. Typically, risk identification includes lessons from previous experience, brainstorming, interviewing key persons and risk review questions to uncover potential risk issues.

Risk analysis is a multi-faceted process that uses techniques already described in Chapter 7 supplemented by other techniques such as Monte Carlo analysis and the use of risk mapping scales and matrices.

Qualitative methods of assessment can be used where risks can be assessed in terms of their probability of occurrence and their impact. The matrix shown in Figure 11.9 provides for low, moderate and high classification boxes to illustrate the extent of the risk (Revill and Gully, 2003).

Figure 11.10 illustrates a radar chart where each risk category is evaluated on a scale of 0–100 with 100 being the highest possible risk. This

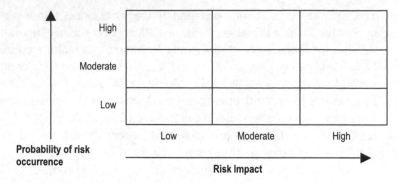

Figure 11.9   Probability/impact matrix.

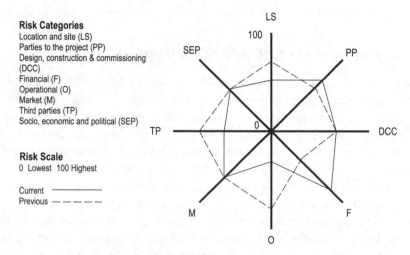

Figure 11.10   Risk radar chart.

representation has the advantage that previous monitoring can be shown to indicate perceived changes in the level of risk for each category. Additional levels of radar chart can be produced to provide risk breakdown within each category.

The evaluation of risk is about comparing the level of risk established from the investigation with what is construed by management to be acceptable. This implies that monitoring should be carried out throughout the project and regular reports should be made to management that clearly articulate the probability of occurrence and the impact of risk.

Summary

The intention of this final chapter is to take a forward view of the main issues facing the efficient provision, upgrading and maintenance of physical infrastructure.

A policy framework is proposed for the development of physical infrastructure, within which policy can be selected and strategy can be developed. A total approach to provision is advocated using a Viable Systems Model in conjunction with the InfORMED System Model, or similar, as described in Chapter 4.

Sustainable infrastructure is seen as a key element of global policy and strategy, whereby growth can be maintained without depleting the Earth's non-replaceable resources, while reducing pollution, especially the release of greenhouse gases into the atmosphere. Conservation is considered to be a vital requirement in the quest to reduce waste and improve efficiency; and actions are proposed to conserve resources.

Attention is drawn to the latest developments in producing energy by new technologies aimed at using replaceable sources to generate power. Special attention has been given to research and development into the production of hydrogen as a fuel from a variety of replaceable and non-replaceable sources that do not cause pollution and are more efficient than existing methods of generating power. The conclusion drawn from an assessment of power generation from replaceable sources is that collectively they have the potential to make a significant contribution to total energy requirements. This implies that infrastructure policies should seek to encourage and coordinate energy production from replaceable sources according to the potential offered by the alternatives described earlier in this chapter.

The concept of service delivery is examined in relation to the growth of PPP as a means of delivering public services from physical infrastructure. Attention is drawn to the need for improvement and the pitfalls associated with conflicts of interest generated by the expectations of private investors.

Relevance is given to the need for innovation in the methods and techniques used to provide sufficient financial resources to meet the increasing demand for physical infrastructure. Suggestions have been made concerning the combined use of national and local infrastructure budgets and the leverage of investment by means of bond issues and loan guarantees, together with tax breaks and other financial incentives.

The chapter concludes with an appraisal of the importance of risk management throughout the whole life cycle of infrastructure projects and the need for close liaison and communication between those assessing risk and project managers.

This book has provided an overview into the imperatives that influence strategies for the global procurement of physical infrastructure. It has attempted to logically develop principles, theories and experience and then relate these to practice with the assistance of case studies at the end of each chapter. The initial chapters have concentrated on defining the nature of infrastructure prior to explaining developing concepts relating to policy and strategy. The realisation of physical infrastructure projects has been covered by chapters concerning procurement and finance. Attention has also been given to the important issue of sustainability, followed by an appraisal of facility management covering operational aspects. The final chapter takes a forward view and makes suggestions and recommendations concerning the future development and provision of physical infrastructure.

References

Beer, S. (1985), '*Diagnosing the system for organisations*', Wiley, New York.

Checkland, P.B. (1981), '*Systems thinking, systems practice*', Wiley, New York.

Conrow, E.H. and Carman, S.L. (2000), '*Risk management experience on hyperion*', Proceedings of the Project Management Institute Annual Seminars and Symposium, September 7–16, Houston, TX, USA.

Federal Highways Administration, FHWA (2004), '*Innovative finance*', Home Page, online at www.fhwa.dot.gov/innovativefinance/

Flood, R.L. and Jackson, M.C. (1988), '*Cybernetics and organisation theory: a critical review*', Cybernetics and Systems, 19, 13–33.

Flood, R.L. and Jackson, M.C. (1991), '*Creative problem solving: total systems intervention*', *Wiley*, New York.

Gray, D. and Tomlinson, G. (2002), '*Hydrogen from coal*', MTR 2002-31, Mitretek Systems, Falls Church, VA.

Howes, R. & Tah, J.H.M. (2003), 'Strategic management applied to international construction', Thomas Telford, London (ISBN 0-7277-3211-0).

Lewis, C.M. and Mody, A. (1998), '*Risk management systems for contingent infrastructure liabilities: applications to improve contract design and monitoring*', Public Policy for the Private Sector, Note 149, August.

NZIER (2004), '*Sustainable infrastructure: a policy framework*', Report to the Ministry of Economic Development, New Zealand.

US Department of Energy (2004), '*Hydrogen posture plan: an integrated research, development and demonstration plan*', The National Academies Committee on Alternatives and Strategies for Future Hydrogen Production and use.

US Environmental Protection Agency (2002), '*The clean water and drinking water infrastructure gap analysis*', EPA-816-R-02-020, online at http://www.epa.gov

US Water Infrastructure Forum (2003), '*Closing the gap: innovative solutions for America's water infrastructure*', online at http://www.epa.gov

Revill, S. and Gully, B. (2003), '*Risk management strategies for future proofing infrastructure projects*', 2nd Annual Utilities and Infrastructure Conference, 10–11 November.

Wall, K. (2003), '*Improved service delivery within the public sector infrastructure environment*', CSIR Building and Construction Technology, Pretoria, South Africa.

Appendix

Tables

Table 1   Present value factors $1/(1 + r)^n$

Years	1%	2%	3%	4%	5%	6%	7%	8%	9%	10%
1	0.9901	0.9804	0.9709	0.9615	0.9524	0.9434	0.9346	0.9259	0.9174	0.9091
2	0.9803	0.9612	0.9426	0.9426	0.9070	0.8900	0.8734	0.8573	0.8417	0.8264
3	0.9706	0.9423	0.9151	0.8890	0.8638	0.8396	0.8163	0.7938	0.7722	0.7513
4	0.9610	0.9238	0.8885	0.8548	0.8227	0.7921	0.7629	0.7350	0.7084	0.6830
5	0.9515	0.9057	0.8626	0.8219	0.7835	0.7473	0.7130	0.6806	0.6499	0.6209
6	0.9420	0.8880	0.8375	0.7903	0.7462	0.7050	0.6663	0.6302	0.5963	0.5645
7	0.9327	0.8706	0.8131	0.7599	0.7107	0.6651	0.6227	0.5835	0.5470	0.5132
8	0.9235	0.8535	0.7894	0.7307	0.6768	0.6274	0.5820	0.5403	0.5019	0.4665
9	0.9143	0.8368	0.7664	0.7026	0.6446	0.5919	0.5439	0.5002	0.4604	0.4241
10	0.9053	0.8203	0.7441	0.6756	0.6139	0.5584	0.5083	0.4632	0.4224	0.3855
11	0.8963	0.8043	0.7224	0.6496	0.5847	0.5268	0.4751	0.4289	0.3875	0.3505
12	0.8874	0.7885	0.7014	0.6246	0.5568	0.4970	0.4440	0.3971	0.3555	0.3186
13	0.8787	0.7730	0.6810	0.6006	0.5303	0.4688	0.4150	0.3677	0.3262	0.2897
14	0.8700	0.7579	0.6611	0.5775	0.5051	0.4423	0.3878	0.3405	0.2992	0.2633
15	0.8613	0.7430	0.6419	0.5553	0.4810	0.4173	0.3624	0.3152	0.2745	0.2394
16	0.8528	0.7284	0.6232	0.5339	0.4581	0.3936	0.3387	0.2919	0.2519	0.2176
17	0.8444	0.7142	0.6050	0.5134	0.4363	0.3714	0.3166	0.2703	0.2311	0.1978
18	0.8360	0.7002	0.5874	0.4936	0.4155	0.3503	0.2959	0.2502	0.2120	0.1799
19	0.8277	0.6864	0.5703	0.4746	0.3957	0.3305	0.2765	0.2317	0.1945	0.1635
20	0.8195	0.6730	0.5537	0.4564	0.3769	0.3118	0.2584	0.2145	0.1784	0.1486
21	0.8114	0.6598	0.5375	0.4388	0.3589	0.2942	0.2415	0.1987	0.1637	0.1351
22	0.8034	0.6468	0.5219	0.4220	0.3418	0.2775	0.2257	0.1839	0.1502	0.1228
23	0.7954	0.6342	0.5067	0.4057	0.3256	0.2618	0.2109	0.1703	0.1378	0.1117
24	0.7876	0.6217	0.4919	0.3901	0.3101	0.2470	0.1971	0.1577	0.1264	0.1015
25	0.7798	0.6095	0.4776	0.3751	0.2953	0.2330	0.1842	0.1460	0.1160	0.0923

Table 2   Present value of an annuity of £1 for N periods: $(1 - (1 + r)^{-n})/r$

No. of payments	1%	2%	3%	4%	5%	10%
1	0.9901	0.9804	0.9709	0.9615	0.9524	0.9091
2	1.9704	1.9416	1.9135	1.8861	1.8594	1.7355
3	2.9410	2.8839	2.8286	2.7751	2.7232	2.4869
4	3.9020	3.8077	3.7171	3.6299	3.5460	3.1699
5	4.8534	4.7135	4.5797	4.4518	4.3295	3.7908
6	5.7955	5.6014	5.4172	5.2421	5.0757	4.3553
7	6.7282	6.4720	6.2303	6.0021	5.7864	4.8684
8	7.6517	7.3255	7.0197	6.7327	6.4632	5.3349
9	8.5660	8.1622	7.7861	7.4353	7.1078	5.7590
10	9.4713	8.9826	8.5302	8.1109	7.7217	6.1446
11	10.3676	9.7868	9.2526	8.7605	8.3064	6.4951
12	11.2551	10.5753	9.9540	9.3851	8.8633	6.8137
13	12.1337	11.3484	10.6350	9.9856	9.3936	7.1034
14	13.0037	12.1062	11.2961	10.5631	9.8986	7.3667
15	13.8651	12.8493	11.9379	11.1184	10.3797	7.6061
16	14.7179	13.5777	12.5611	11.6523	10.8378	7.8237
17	15.5623	14.2919	13.1661	12.1657	11.2741	8.0216
18	16.3983	14.9920	13.7535	12.6593	11.6896	8.2014
19	17.2260	15.6785	14.3238	13.1339	12.0853	8.3649
20	18.0456	16.3514	14.8775	13.5903	12.4622	8.5136
21	18.8570	17.0112	15.4150	14.0292	12.8212	8.6487
22	19.6604	17.6580	15.9369	14.4511	13.1630	8.7715
23	20.4558	18.2922	16.4436	14.8568	13.4886	8.8832
24	21.2434	18.9139	16.9355	15.2470	13.7986	8.9847
25	22.0232	19.5235	17.4131	15.6221	14.0939	9.0770
26	22.7952	20.1210	17.8768	15.9828	14.3752	9.1609
27	23.5596	20.7069	18.3270	16.3296	14.6430	9.2372
28	24.3164	21.2813	18.7641	16.6631	14.8981	9.3066
29	25.0658	21.8444	19.1885	16.9837	15.1411	9.3696
30	25.8077	22.3965	19.6004	17.2920	15.3725	9.4269

Table 3   Area under the normal density function: $\Phi(x) = 1/\sqrt{2\pi} \int\limits_{-x}^{x} e(-t^2/2)dt$

x	0.000	0.010	0.020	0.030	0.040	0.050	0.060	0.070	0.080	0.090
0.0	0.5000	0.5040	0.5080	0.5120	0.5160	0.5199	0.5239	0.5279	0.5319	0.5359
0.1	0.5398	0.5438	0.5478	0.5517	0.5557	0.5596	0.5636	0.5675	0.5714	0.5753
0.2	0.5793	0.5832	0.5871	0.5910	0.5948	0.5987	0.6026	0.6064	0.6103	0.6141
0.3	0.6179	0.6219	0.6255	0.6293	0.6331	0.6368	0.6406	0.6443	0.6480	0.6517
0.4	0.6554	0.6591	0.6628	0.6664	0.6700	0.6736	0.6772	0.6808	0.6844	0.6879
0.5	0.6915	0.6950	0.6985	0.7019	0.7054	0.7088	0.7123	0.7157	0.7190	0.7224
0.6	0.7257	0.7291	0.7324	0.7357	0.7389	0.7422	0.7454	0.7486	0.7517	0.7549
0.7	0.7580	0.7611	0.7642	0.7673	0.7704	0.7764	0.7794	0.7823	0.7852	0.7881
0.8	0.7881	0.7910	0.7939	0.7967	0.7995	0.8023	0.8051	0.8078	0.8106	0.8133
0.9	0.8159	0.8186	0.8212	0.8238	0.8264	0.8289	0.8315	0.8340	0.8365	0.8389
1.0	0.8413	0.8438	0.8461	0.8485	0.8508	0.8531	0.8554	0.8577	0.8599	0.8621
1.1	0.8643	0.8665	0.8686	0.8708	0.8729	0.8749	0.8770	0.8790	0.8810	0.8830
1.2	0.8849	0.8869	0.8888	0.8907	0.8925	0.8944	0.8962	0.8980	0.8997	0.9015
1.3	0.9032	0.9049	0.9066	0.9082	0.9099	0.9115	0.9131	0.9147	0.9162	0.9177
1.4	0.9192	0.9207	0.9222	0.9236	0.9251	0.9265	0.9279	0.9292	0.9306	0.9319

1.5	0.9332	0.9345	0.9357	0.9370	0.9382	0.9394	0.9406	0.9418	0.9429	0.9441
1.6	0.9452	0.9463	0.9474	0.9484	0.9495	0.9505	0.9515	0.9525	0.9535	0.9545
1.7	0.9554	0.9564	0.9573	0.9582	0.9591	0.9599	0.9608	0.9616	0.9625	0.9633
1.8	0.9641	0.9649	0.9656	0.9664	0.9671	0.9678	0.9686	0.9693	0.9699	0.9706
1.9	0.9713	0.9719	0.9726	0.9732	0.9738	0.9744	0.9750	0.9756	0.9761	0.9766
2.0	0.9772	0.9778	0.9783	0.9788	0.9793	0.9798	0.9803	0.9808	0.9812	0.9817
2.1	0.9821	0.9826	0.9830	0.9834	0.9838	0.9842	0.9846	0.9854	0.9850	0.9857
2.2	0.9861	0.9864	0.9868	0.9871	0.9875	0.9878	0.9881	0.9884	0.9887	0.9890
2.3	0.9893	0.9896	0.9898	0.9901	0.9904	0.9906	0.9909	0.9911	0.9913	0.9916
2.4	0.9918	0.9920	0.9922	0.9925	0.9927	0.9929	0.9931	0.9932	0.9934	0.9936
2.5	0.9938	0.9940	0.9941	0.9943	0.9945	0.9946	0.9948	0.9949	0.9951	0.9952
2.6	0.9953	0.9955	0.9956	0.9957	0.9959	0.9960	0.9961	0.9962	0.9963	0.9964
2.7	0.9965	0.9966	0.9967	0.9968	0.9969	0.9970	0.9971	0.9972	0.9973	0.9974
2.8	0.9974	0.9975	0.9976	0.9977	0.9977	0.9978	0.9979	0.9979	0.9980	0.9981
2.9	0.9981	0.9982	0.9982	0.9983	0.9984	0.9984	0.9985	0.9985	0.9986	0.9986
3.0	0.9987	0.9987	0.9987	0.9988	0.9988	0.9989	0.9989	0.9989	0.9990	0.9990

Table 4   Standard normal distribution

z	0.00	0.01	0.02	0.03	0.04	0.05	0.06	0.07	0.08	0.09
0.0	0.0000	0.0040	0.0080	0.0120	0.0160	0.0199	0.0239	0.0279	0.0319	0.0359
0.1	0.0398	0.0438	0.0478	0.0517	0.0557	0.0596	0.0636	0.0675	0.0714	0.0753
0.2	0.0793	0.0832	0.0871	0.0910	0.0948	0.0987	0.1026	0.1064	0.1103	0.1141
0.3	0.1179	0.1217	0.1255	0.1293	0.1331	0.1368	0.1406	0.1443	0.1480	0.1517
0.4	0.1554	0.1591	0.1628	0.1664	0.1700	0.1736	0.1772	0.1808	0.1844	0.1879
0.5	0.1915	0.1950	0.1985	0.2019	0.2054	0.2088	0.2123	0.2157	0.2190	0.2224
0.6	0.2257	0.2291	0.2324	0.2357	0.2389	0.2422	0.2454	0.2486	0.2517	0.2549
0.7	0.2580	0.2611	0.2642	0.2673	0.2704	0.2734	0.2764	0.2794	0.2823	0.2852
0.8	0.2881	0.2910	0.2939	0.2967	0.2995	0.3023	0.3051	0.3078	0.3106	0.3133
0.9	0.3159	0.3186	0.3212	0.3238	0.3264	0.3289	0.3315	0.3340	0.3365	0.3389
1.0	0.3413	0.3438	0.3461	0.3485	0.3508	0.3531	0.3554	0.3577	0.3599	0.3621
1.1	0.3643	0.3665	0.3686	0.3708	0.3729	0.3749	0.3770	0.3790	0.3810	0.3830
1.2	0.3849	0.3869	0.3880	0.3907	0.3925	0.3944	0.3962	0.3980	0.3997	0.4015
1.3	0.4032	0.4049	0.4066	0.4082	0.4099	0.4115	0.4131	0.4147	0.4162	0.4177
1.4	0.4192	0.4207	0.4222	0.4236	0.4251	0.4265	0.4279	0.4292	0.4306	0.4319

1.5	0.4332	0.4345	0.4357	0.4370	0.4382	0.4394	0.4406	0.4418	0.4429	0.4441
1.6	0.4452	0.4463	0.4474	0.4484	0.4495	0.4505	0.4515	0.4525	0.4535	0.4545
1.7	0.4554	0.4564	0.4573	0.4582	0.4591	0.4599	0.4608	0.4616	0.4625	0.4633
1.8	0.4641	0.4649	0.4656	0.4664	0.4671	0.4678	0.4686	0.4693	0.4699	0.4706
1.9	0.4713	0.4719	0.4726	0.4732	0.4738	0.4744	0.4750	0.4756	0.4761	0.4767
2.0	0.4773	0.4778	0.4783	0.4788	0.4793	0.4798	0.4803	0.4808	0.4812	0.4817
2.1	0.4821	0.4826	0.4830	0.4834	0.4838	0.4842	0.4846	0.4850	0.4854	0.4857
2.2	0.4861	0.4864	0.4868	0.4871	0.4875	0.4878	0.4881	0.4884	0.4887	0.4890
2.3	0.4893	0.4896	0.4898	0.4901	0.4904	0.4906	0.4909	0.4911	0.4913	0.4916
2.4	0.4918	0.4920	0.4922	0.4925	0.4927	0.4929	0.4931	0.4932	0.4934	0.4936
2.5	0.4938	0.4940	0.4941	0.4943	0.4945	0.4946	0.4948	0.4949	0.4951	0.4952
2.6	0.4953	0.4955	0.4956	0.4957	0.4959	0.4960	0.4961	0.4962	0.4963	0.4964
2.7	0.4965	0.4966	0.4967	0.4968	0.4969	0.4970	0.4971	0.4972	0.4973	0.4974
2.8	0.4974	0.4975	0.4976	0.4977	0.4977	0.4978	0.4979	0.4979	0.4980	0.4981
2.9	0.4981	0.4982	0.4982	0.4982	0.4984	0.4984	0.4985	0.4985	0.4986	0.4986
3.0	0.4987	0.4987	0.4987	0.4988	0.4988	0.4989	0.4989	0.4989	0.4990	0.4990

Table 5 Random numbers

31	28	18	10	23	22	64	47	55	20	08	73	58	23	68	15	05	98	02	65	49	55	01	44
70	57	67	23	50	96	71	90	97	38	51	52	68	05	48	65	41	20	68	81	75	60	38	49
54	60	16	61	30	58	35	21	64	51	91	16	21	17	39	78	07	66	15	56	23	57	15	57
86	17	71	47	09	08	90	59	04	59	87	02	88	80	64	86	75	54	56	96	11	35	80	97
32	64	43	25	99	16	30	54	73	63	72	22	97	12	60	77	90	86	69	37	70	82	36	48
78	68	39	06	43	90	16	10	28	03	21	60	00	74	78	11	97	74	00	47	91	15	03	63
26	27	47	34	87	28	63	20	04	61	84	49	97	37	46	17	67	77	75	83	67	47	03	18
64	58	24	35	52	31	91	22	83	19	28	50	22	64	54	42	67	60	74	91	17	36	72	10
45	61	19	70	01	76	72	22	82	88	16	72	88	80	97	19	46	73	59	11	01	62	78	16
12	30	97	68	72	81	32	90	04	20	69	31	24	92	52	74	26	47	29	85	77	83	06	70
99	23	92	98	47	50	02	85	04	62	94	57	89	72	59	54	31	75	89	67	17	77	25	00
52	41	03	37	85	91	76	22	55	46	48	17	65	38	70	74	85	71	28	85	23	92	95	04
43	47	35	04	66	95	51	19	36	22	83	09	55	73	06	83	59	23	99	09	06	37	90	05
84	48	60	05	72	10	64	58	00	83	90	76	20	63	39	20	72	19	71	94	34	49	48	68
38	19	91	38	59	54	14	91	13	90	38	99	08	25	46	65	97	16	11	91	75	64	15	76
40	07	09	17	69	02	87	98	10	47	34	18	20	58	71	74	17	32	23	93	92	35	55	87
19	90	69	64	65	51	17	63	35	87	46	02	71	28	27	20	07	92	00	28	05	16	96	10
87	02	22	38	45	62	81	38	30	86	19	71	44	99	47	00	18	38	47	38	47	29	10	77
83	57	82	37	02	47	38	82	47	83	37	29	37	19	28	19	92	72	82	62	28	19	03	72

```
73 28 64 19 37 28 26 28 29 09 00 72 18 36 81 36 46 77 18 23 77 19 01 39 56
84 29 36 44 91 11 39 92 92 63 82 16 37 28 46 28 93 74 92 01 03 95 68 82 99
29 83 80 30 30 82 19 36 63 92 40 92 38 67 72 38 19 12 99 03 00 01 72 89 27
09 58 27 33 84 19 16 68 39 43 13 34 13 82 19 75 50 65 11 15 33 27 43 49 63
02 41 09 60 21 45 62 38 97 42 84 18 42 61 08 83 98 23 09 55 80 51 25 59 46
20 13 59 97 91 68 58 38 18 38 00 03 52 43 93 15 12 18 82 50 06 53 71 15 11

54 55 13 20 70 33 82 28 24 66 04 22 99 66 64 38 05 71 90 08 23 16 33 85 68
57 68 61 37 30 94 81 21 84 81 48 64 45 69 32 98 09 74 59 37 19 06 56 98 02
00 16 45 84 18 33 38 37 39 97 98 76 78 63 98 40 58 73 58 54 21 02 29 62 69
83 28 82 36 91 09 81 24 55 21 57 22 92 50 49 20 35 46 61 97 85 62 08 62 88
95 14 80 68 53 34 79 75 32 54 70 68 46 93 45 04 93 02 84 40 12 33 29 09 14

78 40 29 92 21 20 63 46 16 45 45 41 44 87 26 78 36 57 03 28 77 10 07 89 85
80 62 74 64 26 23 57 99 84 51 29 41 11 66 30 41 40 97 15 72 31 11 42 59 46
23 08 87 23 90 69 65 07 39 85 96 62 74 75 90 70 01 10 86 23 21 88 75 35 97
31 26 65 08 36 08 30 22 55 68 92 06 69 77 16 14 84 34 36 23 43 63 28 36 51
72 70 81 68 17 31 54 16 16 54 09 00 75 02 07 91 93 72 93 16 48 57 27 58 37

90 88 22 92 49 56 85 89 61 84 84 19 32 56 54 85 26 83 79 40 04 30 46 29 24
60 14 25 68 61 37 74 68 79 87 10 67 14 96 92 28 66 44 67 40 79 82 23 53 09
34 56 82 93 29 47 29 47 58 29 02 17 38 46 08 29 93 64 06 99 00 16 12 08 11
82 36 92 10 63 83 63 92 26 27 29 63 91 73 10 10 39 28 84 82 92 62 90 11 88
29 71 83 72 92 01 94 72 39 49 59 92 72 19 00 09 18 38 72 39 03 16 20 69 16
```

Index